Managing Petrochemicals Business

This book offers a comprehensive overview of the contemporary international petrochemicals business and explains related managerial complexities, business challenges, and opportunities to enhance competitiveness. It enables readers to explore critical issues facing the industry, such as profitability and investment economics, optimizing business processes, regulatory aspects, refining different grades of crudes, marketing of refinery products, health, safety, environment, and emerging ESG compliances in petrochemical business.

Features:

- Explains value chain of the petrochemicals industry from a business perspective.
- Highlights economic developments of petrochemicals, applications, marketing, and case studies on business aspects.
- Discusses refinery product mix, pricing, and marketing of refinery products.
- Reviews managerial challenges in the petrochemicals refining business.
- Includes HSE and ESG aspects of the petrochemicals refining business.

This book is aimed at researchers and graduate students in chemical engineering, business and management studies, and petrochemical industries professionals.

Managing Petrochemicals Business

Pramod Paliwal, Sonal Yadav,
and Sudhir Yadav

CRC Press
Taylor & Francis Group
Boca Raton London New York

CRC Press is an imprint of the
Taylor & Francis Group, an **informa** business

Designed Cover Image: Shutterstock

First edition published 2025
by CRC Press
2385 NW Executive Center Drive, Suite 320, Boca Raton FL 33431

and by CRC Press
4 Park Square, Milton Park, Abingdon, Oxon, OX14 4RN

CRC Press is an imprint of Taylor & Francis Group, LLC

Library of Congress Cataloging-in-Publication Data
Names: Paliwal, Pramod author | Yadav, Sonal author | Yadav, Sudhir author
Title: Managing petrochemicals business / Pramod Paliwal, Sonal Yadav and Sudhir Yadav.
Description: First edition. | Boca Raton, FL : CRC Press, 2025. |
Includes bibliographical references and index. |
Identifiers: LCCN 2025001674 (print) | LCCN 2025001675 (ebook) |
ISBN 9781032603117 hardback | ISBN 9781032661919 paperback |
ISBN 9781032661933 ebook
Subjects: LCSH: Petroleum chemicals industry–Management
Classification: LCC HD9579.C32 P35 2025 (print) | LCC HD9579.C32 (ebook) |
DDC 338.4/7661804–dc23/eng/20250411
LC record available at https://lccn.loc.gov/2025001674
LC ebook record available at https://lccn.loc.gov/2025001675

ISBN: 9781032603117 (hbk)
ISBN: 9781032661919 (pbk)
ISBN: 9781032661933 (ebk)

DOI: 10.1201/9781032661933

Typeset in Times
by Newgen Publishing UK

Dedication

To Goddess Saraswati, my parents, wife, and children
—Pramod Paliwal

To my lovely and supportive family
—Sonal Yadav

To my parents and family
—Sudhir Yadav

Contents

Foreword

It is not possible to understand an industry in the fullness of its activity without having a look at how it brands and markets itself-as they say, *"business is marketing"*! For most people outside that industry, the perception of its activity will actually be based on the public-facing communications of individual companies—communications to which branding and marketing are the most prominent contributors. While to those *within* the industry, marketing and branding offer a deeper perspective on the core of their business, very often in the light of those public perceptions that we have just mentioned.

To give an example, due to the sector's specific business model, practitioners in both energy and petrochemicals businesses had traditionally thought that marketing and branding are not relevant to their core activity of producing, distributing, and selling energy. It was thought of as a commodities business, in some ways similar to mining (where large public-oriented marketing campaigns are rare), and a business in which the product—the unit of energy—has an insufficient number of differentiation points to enable neither effective branding nor memorable marketing.

And hence it's a pleasure writing a foreword for the volume titled *Managing Petrochemicals Business* where all aspects of the petrochemicals value chain have been aptly discussed around the pivot of petrochemicals marketing and branding. Starting with the introductory chapter on the global petrochemicals sector, the book has other well-organized chapters on economic contribution of the petrochemicals sector, operations and project management aspects, and health, safety and environment (HSE) and environment social and governance (ESG) issues of the petrochemicals sector—all on the fulcrum of marketing and branding.

As we now know, energy market liberalization, the fight against climate change, new energy technologies, public scrutiny, and a host of other factors have upended this way of thinking. Along with other industrial sectors, both petrochemicals and energy companies have also built powerful and well-capitalized marketing and branding departments, producing eye-catching campaigns and positioning many companies, previously operating under the public radar, into household names. And, in the process, many of those companies have built stronger relationships with their customer base and a better understanding of the market's direction of travel. It could be said that marketing and branding benefit both the company and its customers.

Hence, I am delighted that the authors have included in this book a detailed chapter on marketing and branding within the petrochemical industry. It is an excellent, carefully structured and comprehensive overview that gives more than just a cursory insight into those issues some might erroneously deem peripheral to the actual business of processing oil and gas. As you will see from the case studies that the authors have thoughtfully included in the chapter, marketing and branding in the petrochemical industry include aspects that are essential for both product and business success. Such aspects need to be properly understood, as they are specific to this industry.

On the other hand, a typical petrochemicals business as a largely a B2B operation, will benefit from those B2B marketing and branding strategies that have shown their effectiveness in other sectors. As many businesses have found out the hard way, having your client base consist of other companies, does not absolve you from the responsibility of promoting yourself. If anything, you need to work harder at certain aspects of marketing and branding, as the stakes are often higher than in retail sales. Individual transaction values are more substantial, professional reputations are at stake, and psychologically speaking, in B2B scenarios buyers are a lot less adventurous and willing to try new things than when buying for their own private use.

I am glad that the analysis the authors have included in this volume reflects on such questions, as well as wider issues of sales management, customer relations management (CRM) and contextual B2B marketing tactics. I was particularly intrigued by the discussion of the challenges in petrochemical branding and marketing, of which pricing volatility seems to be the most prominent. To a marketing professional, these paragraphs are a gateway to other parts of the book, as they link to flexible pricing models and hedging strategies covered elsewhere in this work.

In conclusion, the branding of energy and commodities, a once niche question within the wider science of marketing, is now moving center stage, not just for researchers and academics but for businesses at all stages of the energy value chain. In this timely book, the authors have focused their attention on petrochemicals, offering a valuable contribution to the disciplines of marketing and branding, operations and HSE/ESG, and to students and industry practitioners, in the petrochemical sector, energy sector and beyond.

Fridrik Larsen
Author and Professor of Marketing at the University of Iceland, Reykjavik

Preface

Petrochemicals are important to various businesses because they contribute to economic growth, technical advancements, and manufacturing of goods that are essential for comfortable modern day living. Petrochemicals thus play a major role in not only the broader chemical and crude oil industry but also many other industries.

The petrochemical industry has always been a growing area of interest to all concerned: industry, governments, policy makers, and academia. However, with the exception of books that dealt largely with technological issues of the sector, there was an increasing need to have a well-documented volume that deals with the business issues of the petrochemical sector with a global perspective. Also, so far, most of the discussion on petrochemical business has been documented in the form of magazine articles, a few discussion papers, and/or reports by consulting companies. Moreover, as mentioned above, the majority of the work on petrochemical industry has been from a science and technology perspective only. Thus, a gap has always been felt for a book that addresses the vital aspects of petrochemical industry from a business outlook. It's our pleasure that with Taylor & Francis taking the lead, we have been able to tap this potential to write a structured book that deals with economic, managerial and environmental aspects of petrochemical business with an applied technical perspective wherever needed. These include reflections on the petrochemical sector's contribution to economy, petrochemical project management, marketing, health, safety and environment (HSE), environmental, social and governance (ESG), and petrochemical complexes. With well-structured chapters that contain appropriate tables, graphics, and illustrations, the authors have attempted to present a comprehensive view of this business. The authors have been able to do this despite the constraints of book-size, because of well planning and execution that has gone towards this volume. The book also contains a good number of cases that provide varied perspectives on important activities of the petrochemical industry value chain.

Apart from contextual material in the chapters, we have also provided sources of information, so as to enable readers to go for additional readings from databases reports, etc. benefiting those who would like to delve deep into certain specific issues. The book, we feel, should be of significant value to all global petrochemical industry stakeholders, including professionals, business executives, techno-managerial personnel, students, and faculty members of chemical, petrochemical, polymer sciences, and energy sector managerial and technology academic programs.

Dr. Pramod Paliwal
Dr. Sonal Yadav
Dr. Sudhir Yadav

Acknowledgments

Throughout the development of this volume, several extraordinary individuals provided their time and support to keep us motivated, share valuable inputs and help organize field visits. Especially, we thank the following individuals and organizations:

Arjav Pandya
Sanjiv Mehrotra
Shailendra Pandit
Siddhant Saboo
Taylor & Francis India

Former and current students of School of Management, Pandit Deendayal Energy University.

With a deep sense of appreciation, we thank Dr. Gagandeep Singh of Taylor & Francis Books India Pvt. Ltd. for initiating this project. The editorial, publication, and marketing team at Taylor & Francis Books India Pvt. Ltd. cannot be thanked enough. We also appreciate Mr. Praveen Singh Dewal for his constant coordination with the authors.

The authors express their sincere thanks to Ms Jahnavi Vaid for her support and advice throughout the book project.

Further, Dr. Pramod Paliwal and Dr. Sudhir Yadav are extremely thankful to one and all at the School of Management, Pandit Deendayal Energy University, Gandhinagar, Gujarat, for providing the right academic environment and motivation to work on this ambitious project.

Dr. Sonal Yadav expresses her gratitude to the Ahmedabad University fraternity for their motivation and support.

Dr. Pramod Paliwal would like to thank his wife Shally, daughter Gargie, and son Neil for their support throughout the working on this volume and keeping his spirits high.

Dr. Sonal Yadav thanks her spouse, Shashiranjan, and son, Vidur, for endless patience, constant support, and motivation.

Dr. Sudhir Yadav would like to acknowledge the constant support of his wife Sapna and sons Siddharth and Karan towards this book project.

About the Authors

Pramod Paliwal has over 33 years of corporate and academic experience. An MBA and PhD in management, he is a Fellow of the Chartered Institute of Marketing, UK. He has been with School of Management, Pandit Deendayal Energy University, Gandhinagar, Gujarat, India, since 2006. In this profession since 1991, Dr. Paliwal's earlier associations have been with the ACC Limited Pacific Institute of Management, MLS University, India, and the India Offshore Campus of Oxford Brookes University, UK. He has been featured in the 30th Pearl Anniversary Edition of Marquis *Who's Who in the World*, 2013. In 2017, he was selected as Chair of the panel of Indian Studies Abroad by the Indian Council for Cultural Relations (ICCR), MEA, Govt. of India. He is associated with international crowdsourcing consulting firm Wikistrat as a senior analyst in the area of energy. He also speaks and writes on sustainable development and public policy.

Given his academic, professional, and research interest in sustainability, sustainable marketing, and the energy and resources sector, he has made presentations, been a panelist, and conducted workshops at international events spanning India, Europe, Middle East, North Africa, South East Asia, and the USA. He has published on sustainability, marketing, and energy sector issues in refereed international journals.

Sonal Yadav is an assistant professor in eonomics and public policy area, Ahmedabad University, Ahmedabad. She did her BA (Hons.) and MA (Economics) with specialization in economics of human resources from the MS University, Baroda, India. Her MPhil research focused on labour market segmentation in India. She has a PhD degree in economics from Gujarat University, Ahmedabad. Her doctoral thesis examined the impact of economic reforms on industrial labour concerning small-scale engineering industries of Gujarat.

She has presented several research papers at national and international conferences. She has also published research papers in peer-reviewed national and international journals. Her primary research areas include labour economics, agriculture economics, issues of micro, small, and medium enterprises (MSMEs), and gender studies. She has also worked on a research project related to finding the impact of demonetization in the rural economy.

She is a life member of the Indian Economic Association (IEA), Indian Society of Labour Economics (ISLE), Gujarat Economic Association (GEA), Indian Association for Canadian Studies, and WILPF (Women's International League for Peace and Freedom) India chapter.

Sudhir Yadav has more than 34 years of experience, including about seven years in industry and 27 years in academics. He has been with School of Management,

Pandit Deendayal Energy University, Gandhinagar, Gujarat, India, since 2006. As an industry professional, he has worked in marketing and international business divisions of various companies. Dr. Yadav has developed and delivered academic and training modules in the area of oil and gas operations management, oil and gas project management, and oil and gas value chains. His case studies and publications in oil and gas management have been acclaimed by learners. He has conducted in-company and executive management programs for corporate executives. In the past, he has contributed to the natural gas technical skill development initiatives of Petroleum and Natural Gas Regulatory Board (PNGRB). He has visited the USA, Europe, Middle East, and South East Asia for his academic and professional assignments. Dr. Yadav has delivered oil and gas training program in the Middle East. Dr. Yadav has published more than 25 research papers in refereed international journals and two books (published by international publishers).

1 Petrochemical Industry
An Overview

1.1 INTRODUCTION

Chemicals produced from natural gas or crude oil are known as petrochemicals. Petrochemicals are predominantly hydrocarbons, which are compounds consisting of carbon and hydrogen atoms. Petrochemicals are made by a series of processes, like breaking natural gas liquids or refining crude oil.

Basically, the petrochemical sector pertains to the refining, cracking,[1] and polymerization of hydrocarbons—which, as mentioned above, are mostly derived from natural gas and crude oil—into important chemical products.

Petrochemicals are necessary raw materials for the production of a wide range of goods, such as rubber, plastics, synthetic fibers, insecticides, fertilizers, medicines, cosmetics, textiles, electronics, automobiles, construction, healthcare, and a variety of industrial chemicals.

Petrochemicals are important to various businesses because they contribute to economic growth, technical advancement, and manufacturing of goods that are essential for comfortable modern-day living. Petrochemicals thus play a major role in not only the broader chemical industry but also many other industries.

Petrochemicals such as ethylene, propylene, benzene, xylene, methanol, and many more intermediate and end-use chemicals are produced by these processes from basic materials. The industry is important because it can provide raw materials, which are needed to make many common items. Examples of products made from petrochemicals include plastics, fertilizers, solvents, synthetic fibers, adhesives, lubricants, and medications.

The global petrochemical sector is projected to be worth USD 584.5 billion [1] and has a production capacity of 2.6 billion MT [2] by 2022, according to estimates. Major oil and gas companies, chemical manufacturers, and specialized petrochemical firms are some of the key stakeholders in the petrochemical industry.

These companies invest heavily in research and development to innovate new processes, enhance efficiency, and develop sustainable practices.

DOI: 10.1201/9781032661933-1

1.2 CRUDE OIL REFINING AND PETROCHEMICAL PRODUCTS

Crude oil refining involves the separation and conversion of crude oil into various petroleum products, including both fuels and petrochemical feedstocks. Some of the final petrochemical products obtained from the processing of crude oil in a refinery are as follows:

1. **Naphtha**: Naphtha is a light fraction of crude oil and serves as a crucial feedstock (raw material) for the petrochemical industry. It is rich in hydrocarbons such as paraffins,[2] olefins,[3] and aromatics[4] making it suitable for further processing into various petrochemicals, including ethylene, propylene, and benzene.
2. **Gasoline**: Gasoline, also termed as petrol, is a fuel primarily used in internal combustion engines (ICE). It is produced through the distillation and refining of crude oil, where lighter hydrocarbons are separated and processed to meet specific octane and volatility requirements.
3. **Diesel**: Diesel fuel is a heavier distillate fraction of crude oil and is commonly used as fuel for diesel engines in vehicles, trucks, trains, ships, and generators. It is produced through distillation and further refining processes to remove impurities and adjust specifications for different applications.
4. **Aviation Turbine Fuel (ATF)**: ATF, alternatively also as Jet Fuel, is a specialized fuel used in aircraft engines. It is produced through refining processes that yield a high-energy, low-freezing-point fuel suitable for aviation applications.
5. **LPG (Liquefied Petroleum Gas)**: LPG is a blend of propane and butane gases that are liquefied under pressure for storage and transportation. It is generally used as a fuel for heating, cooking, and vehicle propulsion.
6. **Fuel Oil**: Fuel oil is a heavy residual product of crude oil refining and is used as a fuel in industrial boilers, power plants, and marine vessels. It is typically graded according to viscosity and sulfur content, with lighter grades used for heating and heavier grades used for power generation.
7. **Asphalt**: Asphalt, also known as bitumen, is a thick, sticky substance derived from the residual fraction of crude oil refining. It is used primarily in road construction and waterproofing applications due to its binding and waterproofing properties.
8. **Off-Gases**: Off-gases, including refinery off-gas and ethylene cracker off-gas, are by-products of refinery and petrochemical operations containing various hydrocarbons, including ethylene, propylene, methane, and hydrogen. These off-gases can be recycled or processed further to recover valuable olefin streams, which serve as feedstocks for producing additional petrochemicals.
9. **Petrochemical Feedstocks**: Various fractions obtained during crude oil refining, such as naphtha, gas oil, and light hydrocarbons, serve as feedstocks for producing petrochemicals. These include ethylene, propylene, benzene,

toluene, xylene, and other basic building blocks used in the manufacturing of plastics, synthetic fibers, rubbers, resins, and other chemical products.

As discussed above, these were some of the key final petrochemical products derived from the processing of crude oil in a refinery. Each product undergoes specific refining processes to meet quality specifications and market requirements. Several **refined petroleum products obtained from crude oil refining serve as feedstocks for petrochemical production**. These products are typically processed further in petrochemical complexes to produce a varied assortment of chemical compounds used in various industries.

Primary and **secondary** petrochemicals are the two basic groups into which the petrochemical sector can be categorized. *Ethylene, propylene, butadiene, benzene, toluene*, and *xylene* are examples of primary petrochemicals that are made directly from crude oil or natural gas.

These compounds are subsequently processed to produce ensuing petrochemicals such as polyethylene (PE), polypropylene (PP), polystyrene, and PVC (polyvinyl chloride). Olefins and aromatics are another grouping of petrochemicals.

Olefins: The petrochemical industry uses olefins, a family of unsaturated hydrocarbons, extensively as the base material for various goods like rubber, plastics, and synthetic fibers. Olefins that is sometimes also referred to as alkenes, are made from natural gas or crude oil and are essential to the manufacture of numerous common goods. The most widely manufactured olefin is ethylene, which is also used to make PE, the most widely used plastic in the world. Another crucial olefin is propylene, which is utilized to make PP, a material used to make products like carpets, auto components, medical equipment, etc.

Aromatic substances: The production of aromatics involves complex processes, such as catalytic reformation, which converts naphtha into benzene, toluene, and xylenes (BTX). These chemicals then undergo additional processing to produce an array of products. Nylon and polystyrene are two examples of polymers that are made from benzene. Toluene is used in the production of solvents such as paint thinner, nail polish remover, etc. Xylene is used to make popular products like polyester films and fibers. Aromatics' distinctive chemical structure makes them versatile and suitable for a range of applications.

Their aromatic structure, which is *ring-shaped*, gives them an extraordinary stability and resistance to chemical reactions. This characteristic allows them to be easily molded and sculpted without going through chemical changes, which makes them perfect for use in the creation of plastics.

The above section gave an idea about some of the primary refinery products used as feedstocks for petrochemical production. Each feedstock may undergo further processing, treatment, or conversion in petrochemical plants to produce the desired chemical intermediates and end products. After discussing the aspects of crude oil refining and petrochemicals, it would be logical to also discuss the aspects related to natural gas processing and petrochemicals. See Figure 1.1.

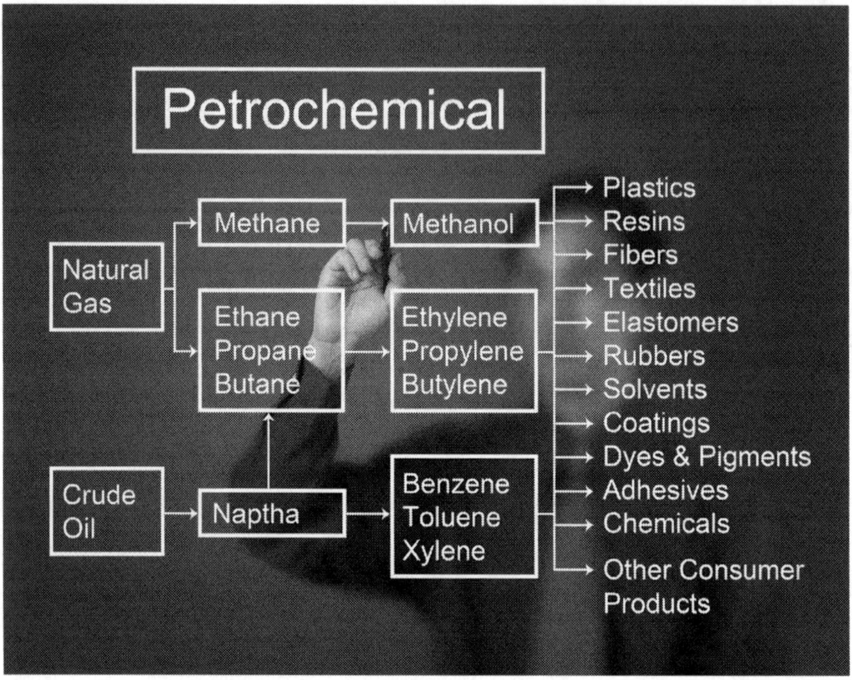

FIGURE 1.1 A Snapshot of Petrochemicals.

1.3 NATURAL GAS PROCESSING AND PETROCHEMICALS

Natural gas processing involves the converting raw natural gas into marketable products. Raw natural gas typically contains various impurities and hydrocarbon gases that need to be removed or separated to meet quality standards and to make the gas suitable for transportation and commercial use. Several petrochemical feedstocks can be obtained from natural gas processing.

Among the important ones are:

1. **Ethane:** Ethane is extracted from natural gas streams and is a necessary raw material for the production of ethylene, one of the petrochemicals used most extensively in the world. Plastics, fibers, and other industrial chemicals are made from ethylene.
2. **Propane:** During processing, another component of natural gas is separated as propane. Another significant petrochemical, propylene, is made from it as a feedstock (raw material). Propylene is used in the manufacturing of plastics, synthetic rubber, and other chemical goods.
   Propane is one of the constituents of LPG.
3. **Butane:** The processing of natural gas can also yield butane, a feedstock (raw material) for the production of butadiene, which is utilized in the

production of synthetic rubber. Butane, like propane, is an important constituent of LPG.

4. **Methane:** While methane is mostly used as a fuel, it may also be processed to create methanol, a chemical feedstock with a wide range of applications in the production of other important common chemicals including formaldehyde, acetic acid, and other polymers.

These feedstocks are essential ingredients for the production of an extensive range of chemicals and materials used in various industries, including plastics, textiles, pharmaceuticals, etc.

1.4 VALUE CHAIN OF THE PETROCHEMICAL INDUSTRY

Now that we have a fundamental understanding of petrochemicals and their relationship to natural gas and crude oil, the following section shall discuss salient aspects of the petrochemical industry's value chain. Each stage in the petrochemical industry's value chain adds value to the raw materials earlier derived from natural gas and petroleum. See Figure 1.2.

An overview of this value chain and its various stages is as follows.

1. **Acquisition of Feedstock:** The petrochemical industry's main feedstocks, crude oil and natural gas, are extracted and refined at the beginning of the petrochemical value chain. Crude oil is processed to produce different *fractions*, such as naphtha, gas oil, and other hydrocarbon products. At the same time methane, ethane, propane, and other gases are obtained during the processing of natural gas in the petrochemical refinery.
2. **Feedstock Conversion:** The next step is conversion of the feedstock. In this stage, fundamental petrochemical ingredients are produced by converting the feedstocks. Important steps in the feedstock conversion consist of the following:
   a. **Steam Cracking:** This process involves breaking down hydrocarbon molecules at high temperatures to produce olefins such as ethylene, propylene, and butadiene.
   b. **Catalytic Reforming:** Used to convert naphtha into aromatic hydrocarbons like BTX.
   c. **Methanol Production:** Natural gas or syngas (a mixture of hydrogen and carbon monoxide) can be converted into methanol, a versatile chemical intermediate.

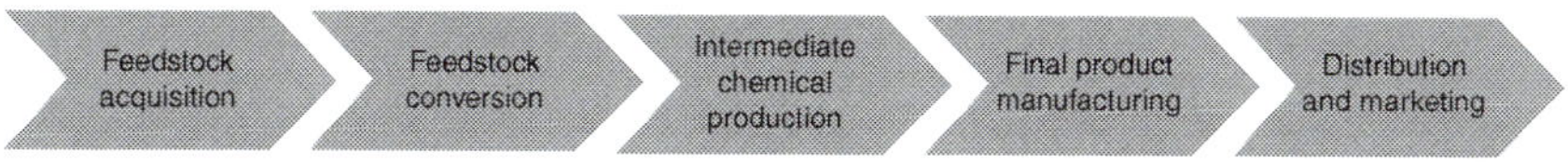

FIGURE 1.2 Petrochemical Industry Value Chain.

d. **Oxidation:** In the process of oxidation, chemicals for a variety of applications are produced through processes such as the oxidation of ethylene to ethylene oxide or propylene to propylene oxide, etc.

3. **Production of Intermediate Chemicals:** Through a series of reactions and purification procedures, the fundamental petrochemical constituents are further processed into intermediate chemicals. A few examples of these are:
   a. **Polymerization:** This is the process by which relatively small molecules called *monomers* are chemically combined to form a very large molecule called a polymer, having the structure of a chain or network. As also mentioned above, olefins like propylene, ethylene, and others can be polymerized to generate polymers like PE, PP, and polystyrene.
   b. **Amination:** In this process, ethylene oxide reacts with ammonia to produce ethylene amines, which are used in the manufacturing of various chemicals and products.
   c. **Oxidation and Hydrogenation:** Processes like oxidation of propylene to acrylonitrile or hydrogenation of benzene to cyclohexane yield important intermediates.
4. **Manufacturing of End Products:** Intermediate chemicals undergo additional processing and formulation to create a variety of end products that are suited to certain uses. A few examples of these are:
   a. **Plastics**: Everyday products like bottles, containers, pipes, and packaging materials are made from PE, PP, PVC, and other polymers.
   b. **Fibers:** Polyester, nylon, and acrylic fibers are manufactured for textiles, carpets, and industrial applications.
   c. **Chemicals:** Various chemicals such as solvents, detergents, fertilizers, pharmaceuticals, and specialty chemicals are produced using petrochemical intermediates.
5. **Distribution and Marketing:** Finished petrochemical products are then supplied to users all over the globe via supply chains. To ensure that petrochemical items are delivered to end consumers in an efficient and effective manner, petrochemical industry professionals ensure that marketing, storage, and transportation activities are undertaken at this stage of the value chain effectively.

   The analysis of the petrochemical industry value chain helps the industry to maximize value creation in the petrochemical business at each stage of this chain. In order to accomplish this the companies engage in research and development (R&D), process optimization, and innovation along the entire value chain. This helps in improving product quality, efficiency, and sustainability for the petrochemical players.

After having undertaken a discussion on the petrochemical industry value chain, it is now contextual to discuss the aspect of petrochemical complexes in the ensuing section.

1.5 PETROCHEMICAL COMPLEXES: AN INTRODUCTION

A petrochemical complex, by definition, is a sizable industrial complex that combines multiple processing units to create a variety of chemical products from hydrocarbon feedstocks like coal, natural gas, and crude oil. By manufacturing the necessary building blocks and intermediates used in the production of products such as plastics, synthetic fibers, rubber, healthcare, fertilizers, and other items, the petrochemical complexes play a significant role in the petrochemical industry. It is worth mentioning that although they are both essential components of the petroleum industry, the petrochemical complexes and crude oil refineries serve different functions and concentrate on different types of products. Petrochemical complexes create chemical products for a range of industrial uses, while refineries produce refined petroleum products largely for automotive energy consumption and other sectors like construction. See Figure 1.3.

The most common petrochemical complexes worldwide are crude oil and natural gas-based. As shall also be discussed in detail in Chapter 3, a crude oil-based petrochemical complex is a facility that uses crude oil and its derivatives as the primary feedstock for the production of petrochemicals. These complexes integrate various processing units to convert crude oil into an extensive range of chemical products, including olefins, aromatics, polymers, and other specialty chemicals.

FIGURE 1.3 Petrochemical Plant.

A natural gas-based petrochemical complex is a large-scale processing facility that uses natural gas as the main feedstock for manufacturing of petrochemicals. In order to convert natural gas into a variety of chemical products, such as *olefins* (e.g. ethylene and propylene), *aromatics* (e.g. benzene and toluene), and other specialized chemicals, the natural gas-based petrochemical complexes have a number of processing units.

Now that we have a general understanding of the petrochemical industry's value chain and are also aware of what petrochemical complexes are, it would be helpful to have a notion of the business processes of the petrochemical industry.

1.6 BUSINESS PROCESSES OF THE PETROCHEMICAL INDUSTRY: AN OVERVIEW

In order to ensure the effective operations and management of petrochemical plants, a variety of business processes are undertaken as per best petrochemical industry practices. Figure 1.4 demonstrates the key business processes carried out in the petrochemical industry.

Business Processes in the Petrochemical Industry are (Figure 1.4):

1. **Feedstock Management:** This process deals with procurement, storing, and managing the raw materials needed for petrochemical production, such as natural gas, crude oil, and other feedstocks. It encompasses tactics for managing inventory levels, sourcing of feedstocks, and streamlining the purchase cycle.
2. **Production Planning and Control (PPC):** All petrochemical facilities need to efficiently plan their schedule and control their production activities with an aim to fulfill demand, optimize resource usage, and minimize downtime. This function also involves forecasting of demand, resource allocation, planned maintenance, and production schedule optimization. Moreover, modern petrochemical complexes also deploy predictive maintenance strategies.
3. **Process Operations and Control**: Process operations and control comprise tracking and managing different operations of different units within the plant, in order to ensure dependable, safe, and efficient manufacturing processes. With an aim of preserving product quality and safety, this activity entails managing process parameters, monitoring equipment performance, and taking corrective actions to deviations, if any.
4. **Product Quality Management:** Like any industry, in the petrochemical business, ensuring product quality is a given thing to meet customer expectations and also comply with regulatory norms. This activity includes quality control procedures (including SQC, statistical quality control), quality testing, and sample and finished product analysis. Moreover, wherever needed, corrective actions are also undertaken to deal with quality issues in this process.

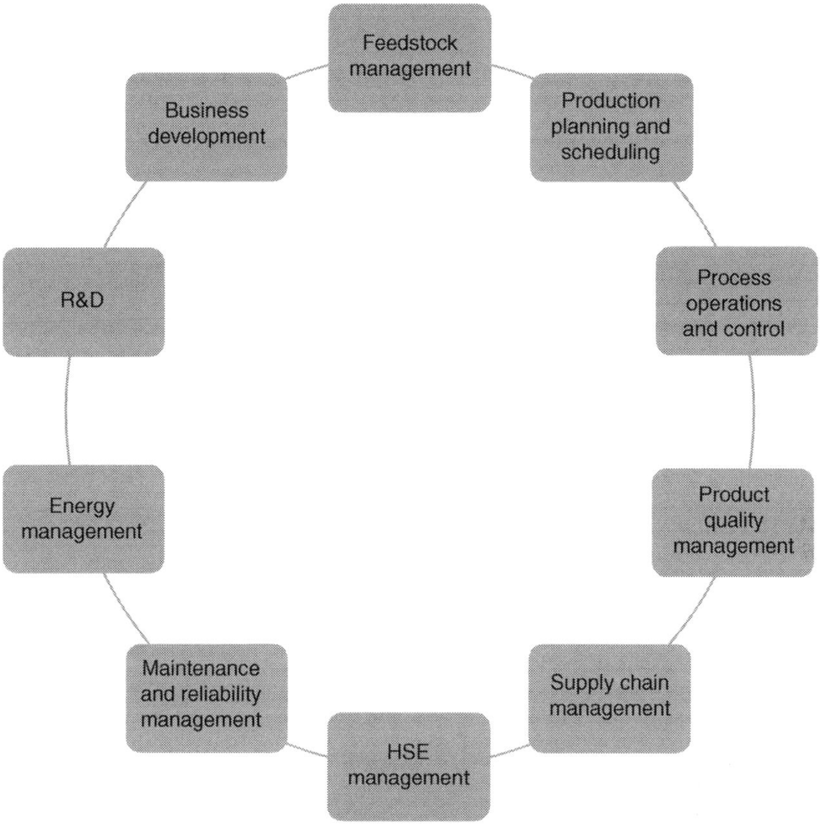

FIGURE 1.4 Petrochemical Industry Business Processes.

5. **Supply Chain Management (SCM):** SCM ensures streamlining the movement of raw materials, intermediates, and final goods along the petrochemical value chain. In order to ensure prompt deliveries, optimizing costs, and minimizing risks of material disruption, this function also involves inventory management, vendor management, logistics, transportation, and warehousing.
6. **Health, Safety, and Environment (HSE) Management:** Worker safety, safeguarding the environment, and adhering to regulatory requirements are prerequisites of the petrochemical sector. HSE management's role thus becomes extremely crucial towards these goals. HSE management deals with setting proper environmental management systems into place, training and development of staff, carrying out risk assessments, designing risk mitigation measures, and also establishing adequate safety management practices.
7. **Maintenance and Reliability Management:** This activity deals with focusing on ensuring the availability and dependability of assets and

equipment for reducing downtime and increasing productivity. This Maintenance and Reliability Management encompasses asset integrity management (AIM), predictive maintenance, preventive maintenance, and other reliability-centered maintenance methods.

8. **Energy Management:** Petrochemical plants should optimize their energy usage and attain energy efficiency in their operations in order to lower their operating expenses and at the same time also lessen the impact on the environment. Energy Management ensures this by deploying energy-saving strategies into practice, maximizing energy use, and also transitioning to alternate energy.
9. **Research and Development (R&D):** Like in any process industry, the petrochemical sector also considers R&D crucial to continuous innovation and keeping pace with technological development. R&D entails exploring novel technologies, new product development, streamlining workflows, and enhancing competitiveness.
10. **Market Analysis and Business Development:** These activities pertain to understanding market trends, appraising client needs, generating new business prospects, and also exploring growing the market share. In addition to this, market analysis and business development activity also considers customer relationship management (CRM), product development, marketing strategies, and market research as its agenda.

All the business processes of the petrochemical sector are thus important for successful and effective operation of petrochemical facilities. After understanding these processes, in the below section, we shall discuss key characteristics of the petrochemical sector.

1.7 KEY CHARACTERISTICS OF PETROCHEMICAL BUSINESS

The uniqueness of the petrochemical industry distinguishes itself from other industrial sectors due to certain special features that the petrochemical industry possesses. We discuss these salient features one by one:

1. **Dependency on Feedstock**: As discussed above, the main feedstocks for petrochemical manufacturing are crude oil and natural gas. Their consistent supply and availability at competitive prices are crucial to the efficient operations of any petrochemical company.
2. **Complex Supply Chain:** Petrochemicals supply chain activities include extraction, refinement, transportation, and processing of raw materials into intermediate and finished products. Effective coordination between all supply chain partners, including distributors, suppliers, and logistics service providers is crucial to manage the supply chain.

3. **Capital Intensive:** The petrochemical sector is capital intensive as its infrastructure, machinery, and technologies require large capital investments. This feature of the petrochemical sector necessitates long-term financial commitments and effective financial management including working capital management.
4. **Technical and R&D Intensive:** Given its complex nature, the petrochemical industry deploys manufacturing methods that are extremely technical, which require a high level of engineering know-how as well as R&D prowess. Augmentation of process efficiency, product quality, and environmental performance for any petrochemical company depends heavily upon continuous innovation and state-of-the-art technology.
5. **Global Market Dynamics:** One of the unique characteristics of the petrochemical industry is the demand-supply dynamics and need for global competitiveness. Thus, it is important that petrochemical companies need to skillfully handle all relevant geopolitical issues, international trade policies, and contextual market trends.
6. **Cyclical Nature:** The petrochemical industry is susceptible to macro-economic variables like GDP growth, industrial activity, and energy costs, all of which have an impact on the cyclical patterns that the petrochemical industry exhibits. While periods of economic prosperity might lead to an increased consumption, economic downturns can result in a decrease in the demand for petrochemical products.
7. **Environmental Concerns and Regulatory Compliances:** Petrochemical firms are subject to a number of regulations that govern the aspects of product quality, health and safety, and environmental protection. In addition to factors like resource conservation, plastic waste, carbon emissions, and other environmental issues are also becoming increasingly significant for the sector.
8. **Globalization and Trade Flows:** Petrochemical manufacturing facilities are dispersed throughout different regions so as to capitalize on feedstock availability, cost competitiveness (due to cheap labor costs such as cheap land, etc.), and market demand. It is known that petrochemical products have a worldwide market. Thus, trade conflicts, taxes and levies, and international trade agreements (restrictive or facilitating) can affect market dynamics and trade flows in the petrochemical sector.
9. **Integration and Diversification:** Petrochemical businesses majorly operate in all three segments of the value chain: upstream, which involves producing crude oil and natural gas; midstream, which involves refining and processing; and downstream, which involves distributing them. Expanding one's business into allied sectors like polymers, specialty chemicals, and refining helps reduce risks and boost competitiveness. Petrochemical companies that operate in all segments of the value chain are called fully integrated petrochemical companies.

10. **Innovation and Sustainability:** To solve environmental issues, lower carbon emissions, and create biobased and renewable alternatives, petrochemical businesses are putting more and more emphasis on innovation and sustainability. Across the board, there is a growing trend of investments in R&D, circular economy projects, and sustainable practices.

These traits influence petrochemical businesses' business decisions and market positioning by defining their operational environment and strategic priorities. The petrochemical industry's size classification will be covered in the parts that follow, along with a quick overview of the global distribution of the industry's major participants.

1.8 PETROCHEMICAL INDUSTRY CLASSIFICATION BASED ON SIZE

The size[5] of the facilities or businesses involved can be used to categorize the petrochemical sector. The following are typical size-based classifications:

1. **Big Petrochemical Businesses:**
   These businesses run sizable, fully integrated petrochemical complexes that generate a variety of chemical products by combining several processing units. They frequently manage different stages of the production process, from feedstock extraction to product distribution, because of their large economies of scale and vertical integration. Multinational companies such as Dow Chemical (USA), Reliance Industries Limited (India), BASF (Germany), ExxonMobil (USA), and SABIC (Saudi Arabia) are a few examples.
2. **Medium-Sized Petrochemical Companies:**
   In comparison to large-scale businesses, medium-sized petrochemical companies run smaller facilities. Because of their specialization in particular goods or specialized markets, they contribute significantly to the petrochemical industry even though companies might not have the same amount of vertical integration or product diversity. These businesses frequently concentrate on local markets or particular industry niches. Some examples of medium-sized petrochemical businesses are Formosa Plastics Corporation (Taiwan), PetroChina (China), INEOS (UK), LyondellBasell (the Netherlands), and Sasol (South Africa).
3. **Small-Scale Petrochemical Companies:**
   Small-scale petrochemical companies operate relatively small facilities that specialize in producing specific chemical products or serving niche markets. These companies may focus on specialty chemicals, fine chemicals, or customized products tailored to meet specific customer requirements. Small-scale petrochemical companies may operate independently or as

subsidiaries of larger companies. Some examples include Braskem Idesa (Mexico), IRPC Public Company Limited (Thailand), AdvanSix (USA), Petronas Chemicals Group Berhad (Malaysia).

4. **Independent Petrochemical Facilities:**
   Independent petrochemical facilities are standalone operations that may not be part of larger integrated complexes or owned by major petrochemical companies. These facilities may focus on producing specific chemical products or providing toll manufacturing services (outsources services) for other companies. Independent facilities often operate on a smaller scale and may have more flexibility in their operations and product offerings. INEOS Styrolution (Germany) Nova Chemicals (Canada), Huntsman Corporation (USA), Westlake Chemical Corporation (USA), and Braskem (Latin America) are a few examples of some leading independent petrochemical facilities across the globe.
5. **Micro-Scale and Specialty Petrochemical Producers**:
   Micro-scale and specialty petrochemical producers operate very small-scale facilities that specialize in producing niche or specialty chemical products. These producers may focus on high-value or low-volume chemicals used in specialized applications such as pharmaceuticals, electronics, or aerospace. Micro-scale manufacturers may apply innovative technologies or practices to meet particular market demands. Elevance Renewable Sciences (USA), Genomatica (USA), LanzaTech (USA), Green Biologics (UK), ONGC petro Additions Limited (India), and Avantium (the Netherlands).

These classifications based on size provide a broad overview of the petrochemical industry's landscape, highlighting the diversity of companies and facilities involved in the production of chemical products from hydrocarbon feedstocks.

A petrochemical company's or facility's size and scope can affect its ability to operate, its market reach, and its strategic placement within the sector.

All things considered, the petrochemical industry plays a crucial role in modern society by supplying the raw materials required for economic growth, technological innovation, and higher living standards while also facing the challenge of striking a balance between expansion and environmental stewardship.

In addition to the ones already covered, Chapter 2 will cover a few more notable players in the global petrochemicals industry, arranged by nation. These players are some of the major contributors to the production and distribution of petrochemical products globally.

However, the petrochemical business is not without its own challenges and constraints. The upcoming section shall discuss some generic constraints faced by petrochemical business, globally.

1.9 CONSTRAINTS OF THE PETROCHEMICAL INDUSTRY

The petrochemical industry is not without its constraints. A number of constraints could impede the petrochemical sector's ability to operate, expand, and develop. Among the principal limitations are:

1. **Availability of Feedstock and Price Volatility:** The petrochemical sector is greatly dependent on the consistent availability of feedstocks, which include natural gas, crude oil, and other hydrocarbons. Any hindrance to continuous availability (which by itself is dependent on many global factors) can be problematic for the industry. Similarly, price fluctuations for feedstocks also can have an impact on industry investment decisions, profit margins, and production costs. Thus, price volatility is also one of the constraints that the industry should overcome.
2. **Compliance With Environmental Regulations:** Petrochemical firms operate under strict environmental regulations about pollution control, waste management, and toxic emissions. Petrochemical companies' requirements to comply with these norms frequently require significant capital (and operational) expenditures on installing environmentally friendly technology and procedures. This may potentially increase the companies' operating costs and subsequently impact their bottom lines.
3. **Energy Costs and Efficiency:** Petrochemical manufacturing processes like steam cracking, hydrogen generation, etc. are energy intensive. Apart from this, the petrochemical firms also confront issues like energy efficiency, greenhouse gas emissions, and rising energy costs. Managing all these issues is a challenge and if not handled adequately, it can result in negatively affecting the companies' profitability and competitiveness.
4. **Demand and Price Dynamics:** Petrochemical products' business is susceptible to many extraneous variables like changing consumer behavior, energy geopolitics, macroeconomic conditions, etc., that can affect the market demand and price of these products. Thus, the bottom-line of petrochemical businesses and their ability to plan outlay and expansion/diversifications can be affected by these dynamics.
5. **Competition Intensity and Business Dynamics:** Worldwide the petrochemical businesses operate under intense competitive rivalry. This can, on the one hand, result in pricing-based competition and on the other hand, the companies are under pressure for technology advancements. These complex dynamics can impact a petrochemical company's profitability and market standing.
6. **Supply Chain Risks and Interdependencies:** The operations of petrochemical companies often rely on complex global supply chains that include numerous suppliers, distribution channels, and logistical networks. The petrochemical business operations and the consistent availability of goods and raw materials can be adversely impacted by any supply chain

interruptions that may be caused due to natural disasters, geopolitical complexities, or any other reasons beyond the control of the companies.

7. **Technological Innovation and Investment:** The petrochemical industry needs constant technological upgradations in order to stay competitive and align with evolving business and regulatory requirements, etc. Thus, the companies need to also constantly provide adequate R&D budgets so as to remain active at this evolving curve.
8. **Geopolitical and Global Regulatory Risks:** Due to their industry's nature the petrochemical firms are constantly exposed to geopolitical and regulation risks. For example, oil and gas, the feedstock for the petrochemical industry itself is prone to global supply chain risks which can be exacerbated by geopolitical tensions. Therefore, the petrochemical industry needs to build in resilience and develop mitigation strategies otherwise its supply chain, investment plans, and market access may get impacted by due to these risks.

After deliberating on the constraints under which the petrochemical industry operates worldwide, and also the measures required to tackle these constraints, we shall attempt to understand the contribution of the petrochemical industry to the larger industrial sector.

1.10 PETROCHEMICAL SECTOR'S SPECIFIC CONTRIBUTION TO VARIOUS INDUSTRIES

The petrochemical industry supports many other sectors in a variety of ways.

1. **Plastics Industry:** The main raw material used to make plastics, which are widely used in modern life, is petrochemicals. Plastics find numerous uses in the construction industry, consumer items, electronics, automotive parts, packaging, and more.
2. **Textiles Industry:** Petrochemicals are utilized in the production of synthetic fibers like polyester, nylon, and acrylic, which are extensively used in the textile industry for clothing, furniture, carpets, and other textile items.
3. **The Pharmaceutical Industry:** Petrochemicals are the building blocks of the pharmaceutical industry, used in the production of excipients, drug delivery systems (DDS), active pharmaceutical ingredients (APIs), and medical devices.
4. **Cosmetics Industry:** The Petrochemical industry makes substantial contribution to the (chemical and pharma based) cosmetics industry. A wide range of skincare, hair care, personal care, etc. products are formulated with substances derived from particular petrochemicals, which include specialty oils, petroleum jelly, waxes, and surfactants (also used in cosmetic soap manufacturing).

FIGURE 1.5 Petrochemical Products' Contribution to Various Industries: An Illustration.

5. **Agriculture Sector:** Modern agriculture sector is closely associated with the petrochemical industry. A range of fertilizers, herbicides, insecticides, and agricultural plastics (pipes/storage tanks/equipment) are made from particular petrochemicals.
6. **Construction Industry:** Apart from basic materials like cement and steel, the construction industry requires a large number of auxiliary products also. Such products which include paints, adhesives, sealants, insulation, roofing materials, pipes, fittings, etc. are manufactured from petrochemicals. Even contemporary cement-concrete uses plasticizers (made from petrochemicals) for long-lasting strength and easy handling.
7. **Automotive Industry:** The automotive industry's reliance on a large number of equipment and components that are made from petrochemicals is a common phenomenon. A range of automotive parts, such as rubber (also used widely in the tire industry), plastics, shade-films, lubricants, adhesives, and gasoline/diesel additives, which improve the safety, performance, and fuel economy of vehicles are manufactured from petrochemicals.
8. **Electronics Industry:** A wide range of electronic gadgets, miniature circuit boards (MCBs), wires, cables, connectors, and insulation materials

are manufactured with materials derived from petrochemicals, including plastics, resins, and specialty chemicals.

9. **Packaging Industry:** Petrochemicals are essential for making plastic films, bottles, containers, and flexible packaging, which offer affordable, strong, and lightweight ways to package and safeguard items while they are being transported and stored.
10. **Energy Industry**: Petrochemicals serve as a source of energy through the production of fuels such as gasoline, diesel, jet fuel, and heating oil, as well as petrochemical feedstocks used in energy generation, refining, and petrochemical processes.

The specific role and contribution of the petrochemical industry to the end-user industry can be better understood by a case. Discussed below is the case and subsequent managerial implications that shall enable us to get a unique perspective, which can be used as a template to identify the important role of the petrochemical industry for various other industries.

1.11 PETROCHEMICAL PRODUCTS' END-USER INDUSTRY: THE CASE OF LEVRAM LIFESCIENCES

Levram Lifesciences (https://levram.co/), an Indian medical device manufacturer, was established in 2015 with a mission of 'making healthcare better'. Levram Lifesciences combines science and technology and its manufacturing prowess for effective and reliable production and delivery of diagnostic and pre-analytical solutions to its customers.

Suitably guided by a professional leadership team, and backed by a professionally qualified and experienced workforce, Levram Lifesciences adheres to the highest quality standards in operations, state-of-the-art manufacturing processes, and in terms of their products. They are an IS0 9001:2015, ISO 13485:2016 certified company, and their products are EU, CE, and WHO GMP certified. The products, processes, and their risk analyses are adherent to ISO 14971:2017. They have a presence in 20 plus countries around the world [5].

For their diverse range of products (as discussed below), Levram's customer base includes hospitals, laboratories, diagnostic centers, research organizations, and medical colleges.

1.11.1 Product-Mix

Levram Lifesciences' diagnostic and pre-analytical solutions-based products primarily address three market segments:

1. Diagnostic labs and/or medical consumables
2. Aesthetic and/or trichology clinics
3. Rapid diagnostic kits

In various combinations, the end products manufactured at Levram Lifesciences are manufactured of the following petrochemical materials:

- Polypropylene (PP)
- Polyethylene terephthalate (PET)
- High-density poly ethylene (HDPE)
- Low-density poly ethylene (LDPE)
- Polyoxymethylene (POM)
- Polystyrene
- Butyl rubber
- Bromo butyl rubber

A brief of the products range manufactured by Levram Lifesciences is as follows.

1.11.2 Diagnostic Labs and/or Medical Consumables

- **Blood Collection Tubes (BCT):** Vacuum and non-vacuum unbreakable, spillage and recollection-free transparent tubes for blood collection with color codes as per ISO 6710:2016. Vacuum blood collection tubes are sterile plastic tubes with a closure that is evacuated to create a vacuum inside the tube facilitating the drawing of a predetermined volume of liquid. Non-vacuum blood collection tubes are sterile plastic tubes with a closure that is not evacuated to draw a predetermined volume of liquid.
  See Figure 1.6.
- **Sample Containers:** For urine, stool, sputum, and saliva collection.
- **Arterial Blood Collection Syringe:** An arterial blood collection syringe is an efficient and advanced medical device, which is pre-coated with balanced lithium heparin.
  See Figure 1.7.
- **Storage Vial:** Used for storing and identifying blood samples such as serum, body fluids, and other biochemical reagents.
- **Micro Centrifuge Tube:** The primary utilization is for centrifugation to test samples in order to separate two immiscible solutions or for precipitation of solute from solvent. These tubes are very useful in storing samples for culture growth, DNA/RNA extraction, amplification, polymerase chain reaction (PCR), blotting experiments, gel electrophoresis, and/or any molecular biology experiment.
- **Centrifuge Tube:** Used to contain liquids during the process of centrifugation, which separates the sample into its components by rapidly rotating it around a fixed axis. These tubes are found in two types, one with a conical bottom and the other self-standing tube with a flat bottom.
- **Swabs:** Swabs that have a push cap system, round bottom, and clear tube visibility are made with high-grade material, and are excellent for

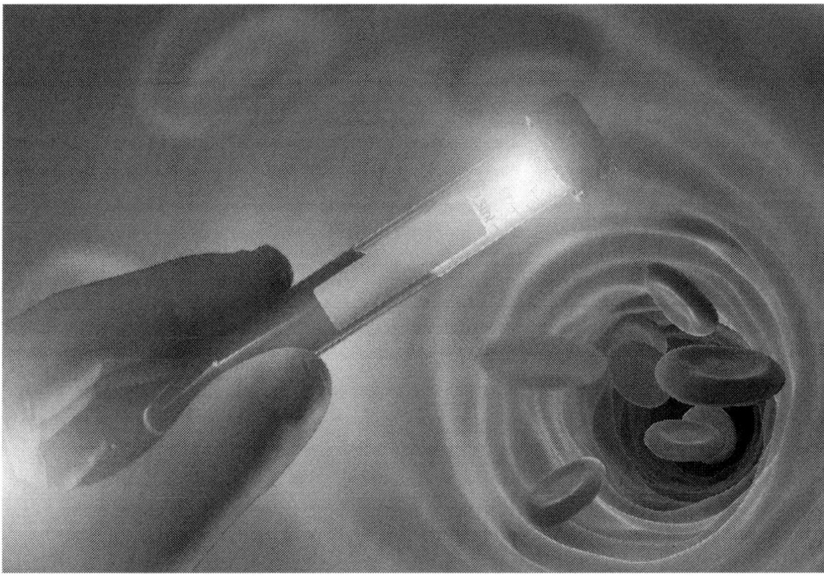

FIGURE 1.6 A Blood Collection Kit.

FIGURE 1.7 Specimen Container.

FIGURE 1.8 Polystyrene Petri Dish.

microbiological sample collection, lab works, and diagnostic purposes such as in gynecology and veterinary.
See Figure 1.8.

- **Pasteur Pipettes:** Also referred to as transfer pipettes, Pasteur pipettes have their stems and bulbs in the form of a single piece made of soft plastic such as PE. The bulb portion is thinner and therefore squeezable. While the pipette portion is thick enough to be rigid.
- **Urine Transfer Device Kit:** These tubes are sterile, vacuum, leakproof, and are made out of clear unbreakable disposable plastic.
- **Petri Dish:** Shallow transparent lidded dishes used for various needs, such as holding growth medium in which cells can be cultures, in microbiology, cell culture, botany and agriculture, entomology, chemistry, for sample storage, contamination detection, and mapping.
- **Hitachi Cup:** Used in most of the popular instruments in diagnostic centers, laboratories, and hospitals, these sample cups are single use and not for autoclave treatment.
- **Safety Needle Holder:** Compatible with various other accessories like needles and safety guards, and the blood collection system.
- **Multi-Sample Needle:** Sterile hypodermic needle commonly known as multi-sample needles are used for blood collection. The uniquely sharpened bevel smoothers the venepuncture method, therefore in comparison with other needles the pain is considerably reduced. It ensures minimal tissue damage and prevents thrombosis. A specially designed rubber sleeve helps in collecting blood in multiple tubes in single venepuncture.

- **Sharps Containers:** Hard plastic container with a wide and clear opening and ergonomically designed handle also has a counterbalanced door for easy and hand-free disposal of sharp instruments such as needles and syringes, safely. It has two kinds of closures—temporary and definite. It also features a secure perimeter lock for safe usage and transportation.
- **ESR Pipette:** ESR (erythrocyte sedimentation rate) pipette eliminates blood transfer, splashing, spraying, and aerosol hazards making it an optimum choice for ESR testing in the laboratory.
- **ELISA Plate:** Clear, sterile ELISA plates allow performing common laboratory procedures on multiple samples at once. These plates have low fluorescence and can be used for ELISA luminescence and fluorescence assays. Designed to identify and quantify protein species, ELISA plates are used for biochemical and cell-based assays.
- **Antigen Extraction Tube:** Developed to meet the needs of antigen-based tests and are easy and safe to use. These extraction tubes give consistent drop sizes while dispensing the buffer into the device.
- **Micropipette Tips:** Micropipette tips are used in laboratories, diagnostic centers, for molecular biology, microbiology, immunology, cell culture, analytical chemistry, to dispense liquid in PCR assays, etc.

1.11.3 Aesthetic and/or Trichology Clinics

- **Regenerative Therapy–PRP:** Platelet-rich plasma (PRP) therapy is a form of regenerative medicine that harnesses and amplifies the natural growth factors found in our blood cells to help heal damaged tissue.
- **Regenerative Therapy–Growth Factor Concentrate (GFC):** GFC kits that cater to a revolutionary technology of using a patient's own blood cells to accelerate healing/aid in regeneration for a variety of conditions from treating skin concerns to hair loss.
- **Regenerative Therapy ACD A:** Anticoagulant citrate dextrose (ACD) solution tubes are intended to be used for collection of whole blood and plasma.
- **ACP Kit:** Acid phosphate total (ACP) test kit is used primarily to diagnose and assess conditions such as cancer of the prostate (diagnose and monitor the effect of the treatment), certain bone diseases (Paget's disease), prostatitis (infection or inflammation of the gland), hyperparathyroidism, and benign hypertrophy of the prostate gland (BPH).
- **Dermaroller:** A *dermaroller* is a skincare device that may help rejuvenate the skin, treat acne scarring, and reduce signs of aging. To use a *dermaroller*, a person simply needs to roll the device on their skin.

  See Figure 1.9.

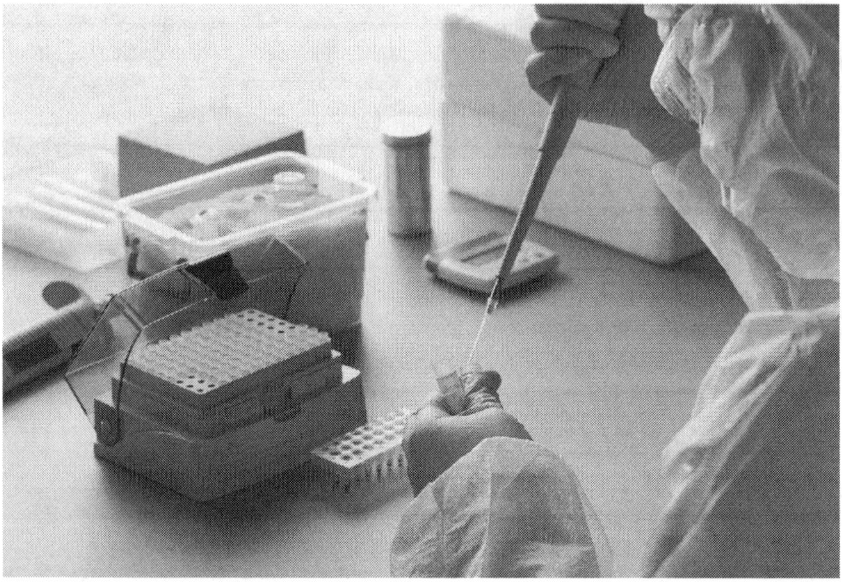

FIGURE 1.9 PRP Kit.

1.11.4 Rapid Diagnostic Kits Segment

- **Diagnostic Kits**: Levram Lifesciences manufactures rapid diagnostic test kits for the detection of pregnancy (hCG), dengue, malaria, HIV, COVID, HCV, HBV, syphilis, leptospira, troponin, typhoid, etc.

1.11.5 Managerial Implications

A medical devices manufacturer that uses materials like PP, polyethylene terephthalate (PET), high-density polyethylene (HDPE), low-density polyethylene (LDPE), polyoxymethylene (POM), polystyrene, butyl rubber, etc. has several key expectations from a petrochemical company. These expectations can be broadly categorized into quality, consistency, compliance, innovation, and support and are discussed as follows:

1. **Quality and Purity**
   a. **High-Purity Grades**: Medical devices require high-purity materials to ensure biocompatibility and patient safety. Manufacturers expect petrochemical companies to supply resins with minimal impurities and consistent quality.
   b. **Traceability**: The ability to trace the origin of materials and their processing history to ensure quality control and regulatory

compliance is a requirement by healthcare industry regulators of many countries.

2. **Consistency**
   a. **Consistent Supply:** Healthcare devices manufacturing is a continuous activity and hence it is important to ensure that disruption in production due to raw material supply delays is avoided. Therefore, dependable and continuous supply chains are essential for Levram Healthcare. At the same time, to take care of this expectation from the buyer, effective supply chain, and logistics management along with reliable delivery schedules, are basic requirements for petrochemical companies (suppliers).
   b. **Uniformity:** Healthcare devices are life-saving devices and hence uniformity during the production process is a crucial requirement. Thus the petrochemical suppliers should also ensure that for the functionality and dependability of medical devices, uniform material qualities supplies {such as melt flow index, tensile strength, and clarity (refraction index) are adhered to.
3. **Regulatory Compliance**
   a. **Standard Compliance:** Medical devices have to be of non-negotiable quality standards like those established by the FDA (Food and Drug Administration), ISO (Quality Certification), and other relevant regulatory organizations. Therefore, it is imperative that the (petrochemical) materials that are used for manufacturing of healthcare devices also have to adhere to similar strict regulations. Major among these regulations are those related to safety, sterilization, and biocompatibility.
   b. **Documentation:** Documentation plays an important role in the healthcare devices industry as an array of products are subjected to clinical trials before being given permission to sell in the market. Thus, the supplier industry (petrochemical) also needs to demonstrate compliance with regulatory standards and provide relevant documentation and certifications to buyers like Levram Healthcare (such as Material Safety Data Sheets and Certificates of Analysis).
4. **Innovation and Development**
   a. **New Material Development:** There are a variety of occasions when healthcare devices manufacturer (buyers) and petrochemical companies (suppliers) work jointly to create new materials that are affordable and capable of providing improved strength, flexibility, or biocompatibility. Healthcare device manufacturers expect such flexibility and transparency from the supplier industry and thus petrochemical companies should also possess such capabilities including qualified scientists and technologists who can undertake such R&D.

   b. **Sustainable Solutions:** Healthcare devices manufacturers are constantly on the radar of regulatory agencies on the aspect of sustainability (e.g. biodegradability of material) of their products during the entire product life-cycle. Thus, development of environmentally friendly materials, such as biodegradable or recyclable polymers, to meet the growing demand for sustainability in medical devices is an important imperative on the part of petrochemical companies also.

5. **Technical Support**
   a. **Expertise and Consultation**: Access to technical support and expertise to help resolve issues related to material selection, processing, and performance.
   b. **Training and Workshops**: Providing training programs and workshops to educate the medical device manufacturer's staff on the properties and processing of new materials.
6. **Customization**
   a. **Tailored Solutions:** Healthcare device manufacturers constantly diversify into new product lines. Thus they should also be completed with suppliers who have the capacity to modify materials (mostly at short notices) to satisfy the needs of various medical devices manufacturers, such as special formulations for particular performance parameters, etc.
   b. **Prototyping Support:** As mentioned earlier, the petrochemical suppliers should possess requisite scientific research capabilities to support the R&D teams of the medical device manufacturers like Levram Healthcare in the prototyping and testing of novel materials.
7. **Economic Considerations**
   a. **Competitive Pricing**: Providing high-quality materials at competitive prices to help medical device manufacturers maintain cost-effective production.
   b. **Long-Term Contracts**: Offering long-term contracts with stable pricing to help manufacturers manage their budgets and production planning.
8. **Sustainability and Environmental Responsibility**
   a. **Eco-Friendly Materials**: Development and supply of materials that are environmentally sustainable, such as those with lower carbon footprints or derived from renewable sources.
   b. **Waste Management**: Collaboration on programs for recycling and managing waste associated with plastic materials.

Petrochemical firms can establish durable partnerships with medical equipment makers and foster innovation in the healthcare sector by fulfilling their expectations. Many of these topics will be covered in more detail in later chapters in relation to the petrochemical products' final customers.

1.11.6 Deduction

The case study details the mutually beneficial partnership between a petrochemical company and a maker of medical devices. All businesses that depend on the petrochemical sector have different expectations from the petrochemical industry, just like the case company. Fulfilling these expectations in a way that benefits both parties. The case study also offers a model for understanding the important role the petrochemical industry plays in the expansion and development of numerous other sectors.

Note: All photographs and information are courtesy of https://levram.co/

Having discussed an end-user industry from the pharmaceutical sector, it would be contextual to also discuss an end-user of petrochemical products from an entirely different sector—the construction materials and packaging industry. The case of Gulanr Plastics is thus discussed in the below section.

1.12 GULNAR PLASTICS PRIVATE LIMITED: CASE STUDY OF AN END-USER OF PETROCHEMICAL PRODUCTS

India-based, Gulnar Plastics Private Limited (www.gulnarplastics.com/) is an end-user business of petrochemical products that manufactures an array of plastic products for various applications. The company's goal is to be the most innovative and preferred woven plastic manufacturer, creating value for all stakeholders. The company intends to serve the building materials and packaging industries globally with cutting-edge and customized products. The company strongly believes in satisfying customers' wants and needs. In partnership with customers, the company constantly innovates, reinvents, and develops new products, thereby enabling them to gain a fair share in the marketplace.

Starting with a brief mention of the company's background below, this case study is aimed at understanding an end-user plastic manufacturing company's perspective that uses petrochemicals as raw materials.

1.12.1 Company Background

Professionals with over 20 years of experience in the plastics sector founded Gulnar Plastics Private Limited in 2010. This group has always been at the forefront of woven plastics technology. After collaborating on multiple occasions with top plastic producers, they made the decision to unite to establish Gulnar Plastics in 2010. The company began with woven sacks and woven fabrics, and after closely observing changes in the plastic sector and working with their clients, they have developed a broad variety of goods for the packaging and construction material industries. The founding motto of Gulnar is "Think Innovations. In order to help clients, partners, employees, and other stakeholders realize their goals, the spirit is to live, eat, and sleep concepts and creative solutions."

The company manufactures a range of products, which include house wraps, roof underlayment, coated woven geomembranes, HDPE geomembranes, and bags for flexible intermediate bulk containers (FIBCs). These products are currently offered in many markets of the world but mainly to companies in the USA, Canada, and Europe. It has established a name for itself in the worldwide production of plastic products that have obtained certification from organizations like the International Code Council (ICC), TRI, LABOR, and ISO by prioritizing the demands of the customer and developing innovative, non-traditional solutions. The plant has implemented quality management systems, environmental performance standards, and safety procedures. It has an ISO certification as well.

The manufacture of PP fabric is now part of its expanded range of capabilities. This business division's primary goal is to produce high-quality PP fabric for usage in the packaging, agricultural, medical, and industrial sectors. Thanks to its superior production abilities and technological advancements, it has emerged as a pioneer in the plastics manufacturing business.

This case study further examines the following aspects:

- Product-Mix of the company.
- The key technologies and manufacturing processes that contribute to Gulnar Plastics' industry-leading position.
- Strategic advantages for Gulnar Plastics.
- The challenges for the company.

These are discussed as follows:

1.12.2 Product-Mix of the Company

The company manufacturers a range of products.

1.12.2.1 Roof Underlayment Solutions

It is made of nonwoven polymers, binders, and tackifiers. Basically, it is woven PP. This product is very strong against tear and lightweight, it has excellent grip in extreme weather conditions, wet or dry. Further it offers stability on steep roofs.

- **Roofer Friendly:** This product was created specifically with roofers in mind. Its amazing traction in both wet and dry conditions was intended to keep roofers on the roof and they offer the best defense against water intrusion. See Figure 1.10.

  Staples are only appropriate if you intend to roof that same day. If staples are exposed to prolonged rain, they will eventually leak. Fasteners ought to be sparingly distributed over the field and concentrated in the horizontal laps.

FIGURE 1.10 Roof Underlayment.

Roof Underlayment is intended to be black. The black surface will quickly dry and reveal chalk lines with clarity, but its minimal bulk will keep the surface much colder than felt soaked with asphalt. The product comes in a 10 square meter roll that weighs only 27 pounds; comes in widths of 18″, 4″, 5″, and 10″, attaches far more quickly and requires fewer fasteners than regular felt.

- **Builder-Friendly:** It can be left outside for up to six months due to its strength and ability to withstand tearing. With today's lighter shingles, Moreoverit lies flat and stays flat—no more creases telegraphing. It is a good substitute for 15 and 30 feet in any kind of roofing.
- **Friendly to Homeowners:** They can serve as a backup barrier against leaks in case of a blow-off. Its exceptional durability and stability last the whole life of the roof. If the roof is exposed, its strong tearing strength will provide reliable coverage.

1.12.2.2 PP Fabrics

PP fabric's strength and resistance to chemicals and abrasion make it suitable for industrial applications, including construction and environmental protection.

- **Unlaminated/Laminated PP/HDPE Tubular Woven Fabric:** The company is involved in the production of both laminated and unlaminated PP/HDPE tubular woven fabric. It produces fabric in sizes ranging from 35 to 140 cm and weights between 50 and 220 GSM. With a lamination thickness of 20 microns, the business can laminate cloth up to 140 cm.
- **Unlaminated/Laminated PP/HDPE Flat Woven Fabric:** The company is involved in the production of both laminated and unlaminated PP/HDPE

flat woven fabrics. It produces fabric in sizes ranging from 35 to 285 cm and weights between 50 and 220 GSM. With a lamination thickness of 20 microns, it can laminate fabric up to 285 cm.

1.12.2.3 Geo Synthetics

- **PP Weaved Bags Geotextiles:** The Company's strength is in weaving geotextiles with the necessary strength in the fabric using high-tenacity PP yarn. The fabric's ability to block sunlight is one of its distinctive qualities. The company produces agro textiles, silt barriers, and fabric with a gauge of GTF 100 to GTF 300.
- **Silt Fence:** To stop sediment-filled runoff, a temporary barrier composed of woven tapes (wire) is called a "silt fence." The primary usage for it is in landscaping. The company provides 24 to 48-inch-wide silt fence cloth in various sizes.
  See Figure 1.11.

1.12.2.4 FIBC Bags

FIBC Bags, sometimes referred to as Jumbo Bags or Bulk Bags, are among the most economical packaging options for the storage and transportation of a variety of goods, including food, metal, chemicals, etc. These bags have a big storage capacity and are made of flat or tubular woven cloth. These bags come in a variety

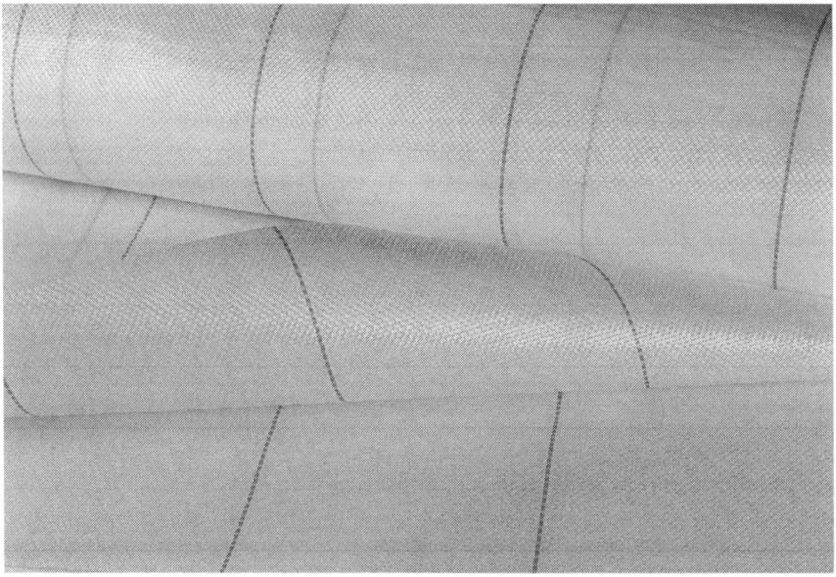

FIGURE 1.11 PP Fabric.

FIGURE 1.12 FIBC Bags Assembly Line.

of styles because they are made to order to satisfy the needs of the customer. For both single-trip (5:1) and multi-trip (6:1 or 8:1) needs, the firm offers bags. The bags have a weight range of 250 kg to 2,000 kg,

See Figure 1.12.

1.12.2.5 Pond Liners

Pond Liners fabric is a long-lasting geomembrane with optimal strength and functionality in mind. It's swiftly gaining traction as the preferred option for situations requiring extreme durability. Heavy-duty liners and covers for a variety of uses in oil fields, landfills, spill containments, remediation, vehicles, railcars, ponds, and pools are the intended uses for these products.

See Figure 1.13.

Product Features and Benefits:

- Construction resistant to scratches, punctures, and chemicals with a high-strength HDPE woven core for exceptional dimensional stability.
- Excellent resistance to oxidation, ozone, and UV.
- The fabric utilized in this product is non-toxic and is made without the use of PVC or any other potentially harmful ingredients.
- Minimal water vapor penetration.
- If submerged in soil, the warranty is for three years; if exposed to the atmosphere, it is for one year.

FIGURE 1.13 Pond Liners.

Applications: This product can be used as pond liners, oil and gas pad liners, liquid containment, irrigation reservoirs and canals, golf course ponds, retention ponds, aquaculture liners, agricultural ponds, and pond liners. It can be utilized for a variety of purposes, including environmental protection, landfill covers and liners, spill containment, remediation, and covers for agricultural, soil, manure, salt, and general-purpose wastewater lagoons.

Products for agriculture applications are ground covers, weed barriers, and crop protection fabrics. PP fabric's durability and UV resistance makes it suitable for protecting crops and controlling weeds, enhancing agricultural productivity.

Application for Medical Products: Surgical masks, gowns, and disposable medical drapes. Nonwoven PP fabric is breathable and lightweight and provides practical barriers against pathogens, which is crucial for medical applications.

1.12.2.6 Flashing Tapes

A self-adhered membrane that can withstand both hot and low temperatures is called flashing tape. It can withstand up to six months of UV exposure.
See Figure 1.14.

Benefits: (i) Custom printing is available, (ii) is resistant six months in the Sun, (iii) the product's Split Release liner allows for quick application and precise tape placement.

1.12.2.7 House Wraps

House wrap is an excellent, reasonably priced way to stop air leaks, manage drafts, enhance the quality of the air within, and shield wall systems from moisture and water damage.

See Figure 1.15.

FIGURE 1.14 Flashing Tape.

FIGURE 1.15 House Wraps.

1.12.2.8 PP Woven Sacks

The company produces PP woven fabrics as well as tubular and flat PP woven bags. It creates FIBC, PP woven geotextile fabrics up to GTF 300, PP spiral tubing, roof under layers, and home wraps. See Figure 1.16.

- **Liner Bags**: 20-micron liner bags are available to meet the needs of the customer.
- **Sand Bags**: Provides sandbags with and without tie strings, in various weights and colors depending on the need.
- **Valve Bags**: Valves with and without lamination, gussets, and valve bags.
- **Sheets:** A variety of sizes and weights are available for sheets. Laminated papers with printing and color are also made.
- **FIBC Bags**: For the secure transportation and storage of bulk goods, FIBC Bags are created and produced in accordance with customer specifications. It provides efficient SCM and logistics. It is composed of strong, rip-resistant fabrics. FIBC Bags minimize the possibility of product damage during transit by ensuring the safe transportation of goods.

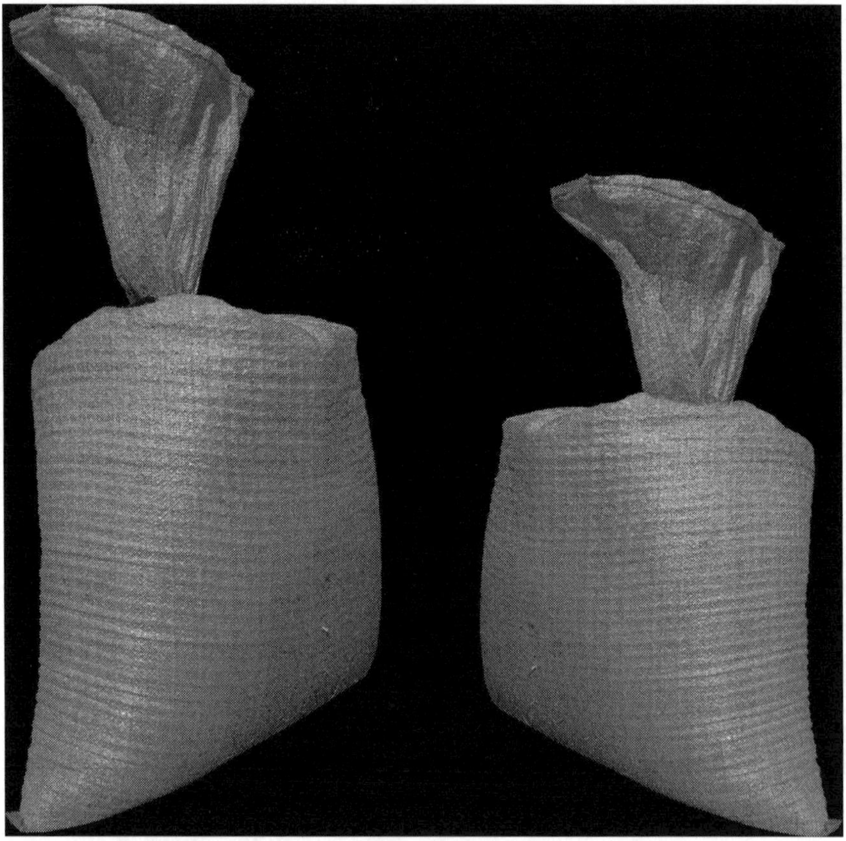

FIGURE 1.16 PP Woven Sacks.

PP fabric is perfect for packing chemicals, building supplies, and agricultural goods since it is robust, lightweight, and moisture-resistant. The company's product line demonstrates the variety of end uses for petrochemical products.

1.12.3 Key Technologies and Manufacturing Processes

The raw materials used by the company to make various end products are various petrochemical derivatives like PE, PVC, PP, etc. The major manufacturing processes required to make various products mentioned in the above sections are briefly discussed in the following paragraphs.

1. **Extrusion:** In this process plastic granules are melted to form a continuous profile. It is an important process to create different plastic products of high quality.
2. **Weaving:** The plastic threads are weaved in a weaving process to produce strong and durable fabric. The fabric used to make woven sacks and geotextiles are from PP.
3. **Coating:** This is the process wherein layers of material are applied on the fabric to increase the durability of fabrics. It helps in making laminated fabrics and geomembranes.
4. **Lamination:** It is used to create composite product by bonding multiple layers of material to enhance the strength and functionality of fabrics and geomembranes.
5. **Quality Control:** Total quality management (TQM) process is followed to meet the quality of various products so that the customers get the product of required specifications. The quality standards are maintained through the use of statistical process control (SPC) across the manufacturing process.

1.12.4 Gulnar Plastics: The Strategic Advantages

The plastic manufacturer has been able to withstand intense competition from the market due to certain advantages that it has been able to create of itself over the years. These are discussed as below:

- **Investment in Technology:** Since the company on a continuous basis invests in the latest manufacturing technologies, it helps ensure high efficiency and high product quality. It also helps company in developing and producing various PP fabrics to meet the varying need of the market.
- **Sustainable Development:** The company is committed to following sustainable manufacturing practices like recycling waste material and making processes energy efficient. Thus this commitment aligns with the global efforts for sustainability and attracts customers who are concerned about the environment. Investment in advanced recycling technology is helping company to process and recycle plastic waste into the production cycle. It has adopted mechanical recycling techniques that convert waste

into reusable raw material. Recycling reduces raw material costs, minimizes environmental impact, and supports sustainable manufacturing practices. Further energy-efficient technologies and practices are followed to minimize the energy consumption. The company uses energy-efficient machinery, optimized production schedules, and energy management systems.

- **Innovative Approach:** The focus of the company's R&D is developing the new applications of PP fabric and improving the quality of existing products. Innovations include enhanced UV-resistant fabrics, biodegradable PP fabrics, and multifunctional textiles. R&D efforts also have a focus on sustainability.
- **Marketing Strategy:** The distribution network of the company is very strong. The company supplies PP fabric and various products to domestic and international markets. Extensive distribution network helps company to maintain stable demand of their products and company growth. One of the company's unique features is its ability to offer customized solutions tailored to specific customer requirements, which sets Gulnar apart from competitors. This flexibility in manufacturing system to offer customized products is very important and offers a competitive edge to meet the specific needs of various industries. The company's logistics and SCM are very efficient. It helps to ensure timely delivery of products to customers across India and abroad. Further the company follows eco-friendly practices, and develops recyclable products to meet the growing demand for sustainable solutions.
- **Superiority in Manufacturing:** The company uses lean manufacturing tools to improve processes, reduce waste and increase the productivity. Tools like Just in Time (JIT), KANBAN, continuous improvement (KAIZEN), Inventory management, etc. are used to optimize various operations. A flexible manufacturing system is helping the company to quickly change to the production line and respond to changing market demands.

1.12.5 Current and Future Challenges for the Company

Gulnar Plastics like any other companies in the plastics manufacturing industry, is highly dependent on its raw material on petrochemical products. The petrochemical industry characteristics leads to certain challenges for the company. These challenges are required to be managed by the company to maintain the competitiveness and sustainability. Some of the primary challenges faced by the company are discussed below.

1. **Petrochemicals Price Volatility:** The basic raw material for petrochemicals are crude oil and natural gas. Since the prices of crude oil and natural gas are volatile in the global market the prices of petrochemical products like PP and PE are volatile. The volatility in crude oil and natural gas may be due to various factors like geopolitical events, demand a-supply situations, economic factors in consuming countries, etc. Thus abrupt increases in

raw material prices erode profit margins as it is challenging to maintain the stable pricing of the end products. The strategy to manage this volatility requires constant price adjustments. The frequent changes in end products prices lead to strained relationships with customers. The company is trying to manage the volatility in raw material prices through strategies like long-term contracts with suppliers and bulk purchasing.

2. **Disruptions in Supply Chain:** One of the challenges is disruption in the supply chain of raw materials. Shortage of critical raw material will cause delayed production schedule and increase in cost. The disruption in supply chain may happen because of trade restrictions, logistics challenges, natural calamities and geopolitical instability.
3. **Regulatory Environment:** In recent times the stakeholders across the world have shown growing concerns for saving the environment. Plastic products which are made from petrochemicals are the targets of environmentalist. The regulatory requirements are getting stronger to reduce the plastic and encourage recycling. To comply with the regulatory requirements companies are required to invest in technologies and facilities for recycling and developing biodegradable products. Regulations also impose penalties on companies for non-compliance and spoil the company image.
4. **Requirements related to Sustainability:** There is increased awareness for environment among consumers and other stakeholders which has driven the demand for environment friendly products/green products. The company to meet this challenge will require investing in research and development to innovate sustainable products and recyclable products. It may require investment in manufacturing processes and sourcing eco-friendly material for their products.
5. **Advancements in Technology:** There is a rapid technological advancement in the field of petrochemicals. Therefore the plastics industries need to continuous innovation to remain competitive. Higher investment in new technologies is required for efficient and sustainable production. If the pace of technological advancement is not matched then it would lead to reduced efficiency, increased production costs. It would further result in lower-quality products.
6. **Market Competition:** The plastics manufacturing industry is highly competitive, with numerous players looking for larger market share. All the players in the industry are regularly investing in the latest technology and sustainable practices. Gulnar Plastics to remain competitive must continually innovate and improve its products to differentiate itself from competitors. So to sustain the market requirements, are ongoing investment in R&D, marketing, and customer service.
7. **International Trade Dynamics:** It refers to import-export tariffs, trade wars between supplier and buyer countries, and international trade policies of both importing and exporting countries. Thus the cost and availability of petrochemical products is affected by the global trade dynamics. Therefore, any changes in trade policies by supplying country or buying country can

lead to increased costs of imported raw materials for end-user industry. It also sometimes limits the access to specific markets. Strategic planning and flexibility are required to adapt to the changes in trade policies.

The company can mitigate these challenges by adopting strategic measures such as (i) diversifying suppliers locally and globally (ii) implementing appropriate financial strategies (iii) investing in sustainable practices, and (iv) focusing on innovation. Implementing these strategies the company will be able to maintain competitive advantage, ensure long-term growth and become a significant player in the industry. Source: www.gulnarplastics.com/

All things considered, the petrochemical industry contributes to economic progress, technological innovation, and improvements in quality of life by offering vital raw materials and components that enable the operation and development of many industries. Given the distinct position of the petrochemical sector within the broader commercial and industrial landscape, it would be appropriate to examine the regulatory needs that are specific to this sector.

1.13 REGULATIONS FOR THE PETROCHEMICAL INDUSTRY

The petrochemical industry is subject to various regulations and standards globally, which may vary depending on the country or region. Some of the important governing rules that guide the petrochemical industry include (See Figure 1.17):

FIGURE 1.17 European Union Regulation REACH—Registration, Evaluation, Authorization, and Restriction of Chemicals.

1. **Environmental Laws:** In order to reduce their negative effects on the environment, governments place laws on emissions, wastewater discharges, and waste management techniques. Environmental impact assessments, hazardous waste management, water pollution controls, and air quality standards are a few examples of these regulations.
2. **Health and Safety Regulations:** Petrochemical industry by the nature of its operations is considered to be a hazardous industry. Thus in order to ensure a safe and healthy workplace for the employees in the petrochemical industry, regulations such as Occupational Health and Safety Series (OHSAS) that deal with safe working conditions, hygienic chemical handling methods, emergency response protocols, worker safety (from accidents), etc. should be followed by the petrochemical manufacturers.
3. **Chemical Safety Regulations:** Chemicals are at the core of the petrochemical sector. Therefore it is important to ensure that they are handled in a safe way in all conditions of usage. Thus in order to evaluate and control the risks connected to the creation, use, and disposal of chemicals—including petrochemicals—regulations like the Registration, Evaluation, Authorization, and Restriction of Chemicals (REACH) [6] in the European Union and comparable laws in other nations are required to be followed by petrochemical manufacturers.
4. **Product Standards and Labeling:** Governments and various other global organizations formulate product standards and labeling regulations to ensure the performance, safety, and quality of goods derived from petrochemicals, such as plastics, chemicals, pharmaceuticals, etc. Therefore as a major supplier industry, the petrochemical sector is also expected to comply with many of such requirements.
5. **Trade Regulations:** An array of trade regulations such as those related to anti-dumping laws, intellectual property, foreign exchange regulations, etc. are also applicable to the petrochemical industry. These regulations may have implications for the sector resulting in affecting market access, trade flows, and the competitiveness of petrochemical goods in various countries. Thus it is important that the sector realizes and interprets all such international trade agreements, tariffs, and import/export rules for their best advantage.
6. **Energy Regulations:** Energy transition and preventing carbon emissions resulting from fossil fuel usage are some of the major issues the petrochemical industry must take cognizance of. Thus the sector must take necessary steps to understand the implications of all such energy-related policies and regulations, especially those pertaining to energy efficiency, renewable energy subsidies, and carbon pricing mechanisms.
7. **Transportation Regulations:** Petrochemical materials are a relatively special category of materials. Thus they need to follow all applicable protocols designed for their safe logistics. All such safety requirements for trucks, railcars, pipelines, and tankers should be adhered to by the petrochemicals industry. In addition to that any rules pertaining to the

handling and labeling of hazardous commodities also need to be taken care of by the industry.

8. **Land Use and Planning Regulations:** The site, building, and operation of petrochemical facilities, such as refineries, chemical plants, and storage terminals, may be impacted by zoning laws, land use regulations, and permitting procedures.

   These are only a handful of the laws that are in place around the world to regulate the petrochemical sector. A given nation or region's specific regulations may differ depending on local laws, regulatory bodies, industry standards, and stakeholder goals. The final section of the introduction covers the petrochemical industry's impact on society as a whole.

1.14 CONTRIBUTION OF PETROCHEMICAL SECTOR TO THE SOCIETY

The petrochemical industry is important to society in a number of ways (see Figure 1.18).

1. **Economic Contribution:** Petrochemicals are essential to many industries, such as construction, plastics, textiles, medicines, and cosmetics. They supply the raw resources needed to make a wide range of goods, which promotes investment opportunities, job growth, and economic expansion.

FIGURE 1.18 Petrochemical Sector Makes a Vital Contribution to the Society.

2. **Technological Innovation:** The petrochemical sector leads the way in this area, continuously creating new materials, methods, and tools. This invention promotes advancement across a range of industries by raising the caliber, effectiveness, and efficiency of products.
3. **Energy Production:** Petrochemicals are an essential energy source that power industrial activities, heating, transportation, and the production of electricity. They are essential to supplying the world's energy needs and guaranteeing energy security.
4. **Infrastructure Development:** Building infrastructure, such as roads, bridges, pipelines, and buildings, requires petrochemicals. They offer strong, reasonably priced materials that are essential for today's building projects.
5. **Healthcare and Pharmaceuticals:** Petrochemicals are the fundamental components of many necessary medications and medical equipment, advancing healthcare and raising standards of living for countless numbers of people globally.
6. **Agriculture and Food Production:** The production of packaging materials, insecticides, and fertilizers requires petrochemicals, which promotes food security and agricultural productivity.
7. **Environmental Conservation:** Although the petrochemical sector has an adverse effect on the environment, it also helps to conserve the environment by developing sustainable practices, recycling programs, and cleaner technology.

All things considered, the petrochemical industry contributes to economic growth, technological advancement, and enhancements to many facets of human existence, playing a complex role in society. But it also has to deal with criticism and issues around environmental sustainability, thus in order to lessen its negative effects, responsible behavior and ongoing innovation are required.

1.15 CONCLUSION

Petrochemical business is vital to the growth of several industrial sectors. The basics of this industry discussed in this chapter shall be helpful to further understand the industry from many other perspectives economic, operations, technology, marketing, health environment and safety—in the following chapters.

NOTES

1 Heat, pressure, and occasionally catalysts are used in the process of "cracking," which breaks apart heavy hydrocarbon molecules into lighter ones.

2 Paraffins are a mixture of hydrocarbons and are combustible in nature. Natural gas and petroleum contain a large amount of paraffin. Depending upon carbon atoms per molecule, paraffins are gaseous, liquids, or solids.

3 A class of compounds known as olefins is composed of carbon and hydrogen atoms joined by one or more double bonds between carbon atom pairs. They serve as the raw materials for goods including adhesives, detergents, and plastics. Olefins include propylene, ethylene, and 1, 3-butadiene [3].

4 Aromatics are compounds generated from crude oil and, in smaller amounts, from coal. They are named for their characteristic scented fragrance. Hydrocarbons, or organic compounds made entirely of carbon and hydrogen, are what are known as aromatics. Without these components, life on Earth would not be conceivable. The primary aromatics are toluene, xylenes, and benzene; they serve as raw materials for a variety of consumer goods. The aromatics sector produces a wide range of products that are used on a daily basis and advantages, including lightweight design, safety, comfort, and durability. Products for a wide range of industries, including sports, fashion, telecommunications, medical, and hygiene, are made with aromatics [4].

5 The size-based classification is based on information available in the public domain and has been mentioned from an academic perspective only. This in no way attempts to compare these companies with other companies. Each company is uniquely placed and contributes significantly to the petrochemical industry worldwide. Also, the terms large, medium, small, micro, independent, etc. are not absolute, but contextual in nature.

REFERENCES

1. Statista. (n.d.). Petrochemical market value worldwide 2018–2022. Statista. Retrieved May 23, 2024, from www.statista.com/statistics/696921/petrochemical-market-value-worldwide/
2. Statista. (n.d.). Petrochemical market value worldwide 2018–2022. Statista. Retrieved May 23, 2024, from www.statista.com/statistics/1260037/global-petrochemical-production-capacity/
3. Chemical Safety Facts. (n.d.). Olefins. Chemical Safety Facts. Retrieved May 23, 2024, from www.chemicalsafetyfacts.org/chemicals/olefins/
4. Britannica (n.d.). Aromatic Compound. Britannica.com. Retrieved March 3, 2025, from https://www.britannica.com/science/aromatic-compound
5. Levram Lifesciences. (n.d.). About Us. Retrieved May 24, 2024 from https://levram.co/about-us/
6. European Commission. (n.d.). REACH regulation. European Commission. Retrieved May 23, 2024, from https://environment.ec.europa.eu/topics/chemicals/reach-regulation_en

2 Petrochemicals and Economic Development

2.1 INTRODUCTION

The petrochemical industry contributes to economic development by increasing gross domestic product (GDP), creating employment opportunities, and generating export earnings. As discussed in Chapter 1, petrochemicals are derived from various chemical compounds that are by-products of crude oil refining. Natural gas and crude oil serve as the primary raw materials for the production of petrochemicals. Therefore, the profitability of a petrochemical business is closely tied to the prices of crude oil and gas. The volatility in crude oil prices and supply impacts petrochemicals' production costs and pricing strategy.

As per the International Energy Agency (IEA) report,[1] petrochemicals constitute about 14 percent and 8 percent of total oil and gas demand, respectively. As the global industry develops, the petrochemical market is set to expand further. Petrochemicals are now the primary driver of oil demand, surpassing aviation and shipping. By 2030, the sector will account for more than one-third of the growth in oil demand and nearly half by 2050. The sector also significantly contributes to the growth in gas demand, accounting for 7 percent of the approximately 850 billion cubic meters global increase by 2030 (IEA) [1].

Petrochemicals are crucial in numerous industries, including agriculture, construction, automotive, electronics, plastics, packaging, textiles, and personal care. They are also found in everyday household items, medical appliances, and parts of modern energy systems. Thus, the petrochemical industry has enriched human life across countries in a unique way. It is expected that the demand for petrochemical products will increase as these industries grow.

However, the growing inclination of consumers towards environment-friendly products may hamper market growth in the coming years. As a result, petrochemical companies must carefully manage their operations and investments to ensure long-term viability. This will encourage industries to actively innovate and invest in new technologies to develop sustainable solutions and provide environmentally friendly consumer products.

DOI: 10.1201/9781032661933-2

2.2 PETROCHEMICAL INDUSTRY AND ECONOMIC DEVELOPMENT

The petrochemical industry is crucial for the economic development of many countries. The industry's location is influenced by the availability of raw materials, market dynamics, and economic policies of different countries. Recognizing these regional strengths can help companies strategically position themselves in the global petrochemical market and take advantage of growth opportunities. See Figure 2.1.

The petrochemical industries in Asia Pacific, North America, and Middle Eastern countries, including Korea, Taiwan, and more recently, China, have emerged as leaders in the petrochemical sector, surpassing the United States and Europe. These countries have experienced long-term growth due to their access to raw materials, low production costs, technological advancements, market demand, and an increasingly skilled workforce. All these factors have significantly contributed to the development of the petrochemical manufacturing hub. The industry has played a vital role in creating jobs, generating revenue, and contributing to exports, thereby driving economic development in these countries.

Let us discuss the petrochemical industry's geographical spread and competitive advantages and how it contributed to economic development in these regions.

FIGURE 2.1 Petrochemical Industry Contributes to Global Economic Development.

2.2.1 North America

North America is an advantageous region for chemical production due to the plentiful supply of natural gas liquids (NGLs) resulting from the discovery of shale gas in the United States and Canada. North America leads in ethane-based petrochemicals, attracting numerous global chemical companies to expand or build facilities in this region (S&P Global's Chemical Economics Handbook[2]) [2]. The increase in demand for plastic, fertilizer, synthetic rubber, and detergents has led to the fast growth of the petrochemical market. Investment in technological innovation has enhanced productivity in petrochemical industries.

United States: The United States is a major player in the petrochemical industry, with significant production capacities located along the Gulf Coast in states like Texas and Louisiana. This region benefits from abundant natural gas supplies, extensive infrastructure, and proximity to domestic and international markets. The country is a significant exporter of petrochemical products to India, China, and Europe.

Canada: Canada has a significant petrochemical industry, especially in Alberta, which has large reserves of natural gas and crude oil. The country is a major producer of unconventional oil. The Canadian industry benefits from access to feedstocks at market-related prices and the presence of world-scale modern plants, giving it a strong competitive advantage. Canada's standard of living significantly depends on the domestic production of petrochemicals.

Competitive Advantages

- **Access to Shale Gas:** The United States and Canada benefit from abundant reserves of shale gas, which provides low-cost ethane and other NGLs. These serve as crucial feedstock for petrochemical production, helping to reduce production costs and enhance the competitive edge of petrochemical producers.
- **Technological Innovation:** North America, especially the United States, leads in technological advancements, such as digitalization, process optimization, and advanced catalysis. Substantial investment in research and development (R&D) strengthens the region's competitiveness.
- **Well-Developed Infrastructure:** Infrastructure in the region is well-developed, with extensive pipeline networks, efficient transportation systems, and proximity to major consumer markets facilitating the movement of raw materials and finished products.
- **Regulatory Environment:** Although environmental regulations are strict, North American companies excel in compliance, often driving innovation in sustainable practices and technologies.

2.2.2 Europe

Europe is prioritizing sustainable practices and the development of bio-based chemicals. Strict environmental regulations are driving innovation in sustainable practices and technologies, positioning Europe as a leader in green chemistry and sustainable development. Major European countries involved in petrochemical production include the UK, France, Germany, the Netherlands, Italy, Spain, Poland, and Belgium.

Western Europe: Countries such as Germany, the Netherlands, and Belgium have significant petrochemical industries. Germany is the leading producer of petrochemicals in Europe, closely followed by the UK and France. These countries enjoy the advantages of a well-developed industrial base, advanced technology, and strong transportation networks.

Eastern Europe: Russia has a significant petrochemical sector due to its extensive natural gas and crude oil reserves. Poland and Hungary also play notable roles in this region, with significant contributions from Western Siberia, Volga-Urals, and the Russian Far East.

Competitive Advantages

- **Technological Advancement:** Europe is recognized for its advanced chemical engineering, process technologies, and innovation in specialty chemicals. The region's companies lead in developing high-performance materials and sustainable solutions.
- **High-Value Products:** European petrochemical companies often focus on producing high-value specialty chemicals and advanced materials, which result in higher margins and distinguish them from competitors.
- **Sustainability Initiatives:** Europe is leading the way in sustainability initiatives such as the circular economy, recycling, and bio-based chemicals. These initiatives not only adhere to strict environmental regulations but also resonate with eco-conscious consumers. This shift is impacting innovation, regulatory compliance, and market offerings in the sector.
- **Strong Regulatory Framework:** The regulatory environment in Europe is stringent, compelling companies to innovate and embrace top safety, efficiency, and environmental protection practices.

2.2.3 Middle East

The strategic location of the region, along with its export-oriented petrochemical industries, contributes to economic growth. The region benefits from abundant natural gas and crude oil reserves, providing a significant cost advantage in feedstock. Investments in petrochemical infrastructure and capacity expansion lead to economic development and job creation in this region.

Saudi Arabia: Saudi Arabia is one of the largest producers of petrochemicals, thanks to its substantial reserves of crude oil and natural gas. The country is a prominent member of the Organization of the Petroleum Exporting Countries (OPEC). Major oil fields in the country include Ghawar, Safaniya, and Khurais. Additionally, petrochemical complexes such as those in Jubail and Yanbu are among the largest in the world.

Qatar, the United Arab Emirates, and Iran: These countries are known for their significant petrochemical industries, which make use of their large natural gas reserves.

Competitive Advantages

- **Abundant Hydrocarbon Reserves:** Countries like Saudi Arabia, Qatar, and the UAE possess plentiful reserves of natural gas and crude oil, which gives them a significant cost advantage for obtaining feedstock. These resources are crucial for producing a wide range of petrochemical products.
- **Economies of Scale:** The Middle East is home to some of the largest petrochemical complexes globally, resulting in significant economies of scale that reduce per-unit costs and enhance competitiveness.
- **Strategic Geographic Location:** The region's proximity to major shipping routes and markets in Asia, Europe, and Africa makes it easier to export petrochemical products globally. Participation in regional trade agreements enhances market access, reduces trade barriers, and promotes exports.
- **Investment in Expansion:** Middle Eastern nations are making significant investments to expand their petrochemical capacities and diversify their economies, strengthening the region's status as a leading petrochemical hub.

2.2.4 Asia Pacific

This region is presently the major producer and consumer of chemicals. The strong economic growth in China and India is leading to increased demand for consumer goods and services. This region is a global manufacturing hub, producing a diverse range of goods from electronics to textiles, which in turn drives the demand for petrochemical products.

China: China has rapidly expanded its petrochemical industry to meet the demands of its growing economy. Major petrochemical hubs are located in northeast and offshore regions, such as Guangdong and Zhejiang provinces. China's petrochemical sector is highly advanced, and its global manufacturing capacity is rapidly increasing. By the end of 2024, China's petrochemical capacity will account for nearly a quarter of the world's total. (Bloomberg[3]) [3].

India: India is one of the fastest-growing markets for petrochemical products. The Indian petrochemical industry has experienced significant growth over the past few decades and has substantially contributed to India's economic development through significant investments in capacity expansions. The petrochemical industry in India is dominated by large public sector companies, which have heavily invested in developing infrastructure and technology to increase production capacity and improve product quality.

There is a concentration of petrochemical industries in Gujarat, Maharashtra, and Tamil Nadu. Coastal regions have a higher agglomeration of refineries and plants than landlocked regions.

Japan and South Korea: Both countries have advanced petrochemical industries, focusing on high-value specialty chemicals and polymers. Japan's petrochemical industry uses naphtha from both domestic and overseas oil refineries as a raw material to produce a wide range of chemical products.

Competitive Advantages

- **Growing Domestic Markets:** Rapid industrialization, urbanization, and rising consumer incomes in emerging economies like China and India drive strong demand for petrochemical products. The rising population and growing sectors such as construction, automobile, and packaging boost the consumption of petrochemical products across various applications. This domestic market growth supports regional production and investment in petrochemical products.
- **Strategic Investments:** In order to meet the growing demand, the company is making significant investments in new petrochemical plants and infrastructure. This includes transportation networks, ports, and logistics. These investments will increase production capacity and technological capabilities, leading to cost savings. Additionally, substantial investments in research and development will improve current processes and facilitate the development of new products.
- **Integrated Value Chains:** China is developing integrated petrochemical value chains that encompass the entire process from raw material extraction to the production of high-value specialty chemicals. This integration is aimed at improving efficiency and cost-effectiveness. Lower labor costs, compared to those in developed countries, help to reduce production costs and enhance competition.
- **Government Support:** To promote growth and attract investment in the petrochemical sector, governments often provide policy support such as tax incentives, subsidies, and infrastructure development. The establishment of special economic zones (SEZs) and industrial clusters dedicated to petrochemicals fosters the sector's growth.

2.2.5 Latin America

Economic development and industrialization in Brazil and Argentina support the growth of petrochemical industries. The agriculture sector also heavily relies on petrochemicals for fertilizers and pesticides. Investment in infrastructure projects and the growth of industrial sectors, including agriculture, manufacturing, and construction, are vital for improving the quality of life.

Brazil: Brazil's petrochemical industry is well-developed, especially in the southern region, like in São Paulo. The country benefits from both domestic oil production and imports.

Mexico: Mexico's petrochemical facilities, concentrated around the Gulf of Mexico, take advantage of its proximity to the United States and access to oil and gas reserves.

Competitive Advantages

- **Resource Availability:** Countries such as Brazil and Argentina possess substantial reserves of natural resources, which serve as raw materials for petrochemical production.
- **Regional Market Access:** Latin America's petrochemical market is expanding due to economic growth and trade agreements, which facilitate market access.
- **Cost Competitiveness:** Some Latin American countries benefit from lower labor and operational costs than other regions, enhancing their competitive position in the global market.
- **Government Policies:** Supportive government policies, such as investment incentives and infrastructure development, contribute to the region's petrochemical industry growth.

2.2.6 Africa

Economic growth and job creation in Africa are driven by the expansion of petrochemical industries. Countries like Nigeria and Egypt leverage their natural resource endowments to develop their petrochemical industries.

Nigeria: Nigeria's petrochemical sector is growing due to its extensive oil and gas reserves. The country is focused on improving its refining and petrochemical capabilities.

Egypt and South Africa: These countries have well-established petrochemical sectors. South Africa is particularly notable for its production of synthetic fuels and chemicals.

Competitive Advantages

- **Natural Resource Endowment:** Nigeria and Egypt possess substantial crude oil and natural gas reserves, providing a strong raw material base for the petrochemical industry.
- **Emerging Markets:** The African continent represents an emerging market with growing demand for petrochemical products due to population growth and economic development.

- **Investment Opportunities:** There is significant potential for investment in petrochemical infrastructure and capacity expansion, driven by local demand and export opportunities.
- **Strategic Location:** Proximity to Europe and Asia via key maritime routes offers logistical advantages for exporting petrochemical products.

2.3 PRIMARY PETROCHEMICALS

The petrochemical industry is affected by globalization. The demand for chemicals and plastics is influenced by worldwide economic conditions, which are directly linked to the demand for consumer goods. Key building blocks for chemicals and plastics include ethylene and propylene. They are used to produce various products such as polyethylene (used in packaging material), polypropylene (used in automotive parts, textiles, and consumer products), polyvinyl chloride (used in construction for pipes, window frames, and flooring), polystyrene (used in packaging, disposable cups, and insulation materials).

Olefins, aromatics, and methanol are considered primary petrochemicals and serve as precursors for various chemical products. The analysis primarily focuses on the basic petrochemicals such as ethylene, propylene, butadiene, benzene, toluene, xylenes, and methanol. See Figure 2.2.

The major consumers of primary petrochemicals[4] are Northeast Asia (largely China), North America (mainly the United States), the Middle East, and Western Europe. See Table 2.1.

2.4 PETROCHEMICAL FEEDSTOCK: MARKET SCOPE

In its growing stage, the petrochemical industry used coal and ethanol as primary feeds; however, over the years, with the advent of technology, the feedstock shifted from coal and ethanol to crude-derived naphtha and natural gas components such as ethane and propane. In 2021, the largest petrochemical feedstock used worldwide was naphtha, followed by natural gas, coal, and ethane (S&P Global's Chemical Economics Handbook: Petrochemical Feedstocks) [2].

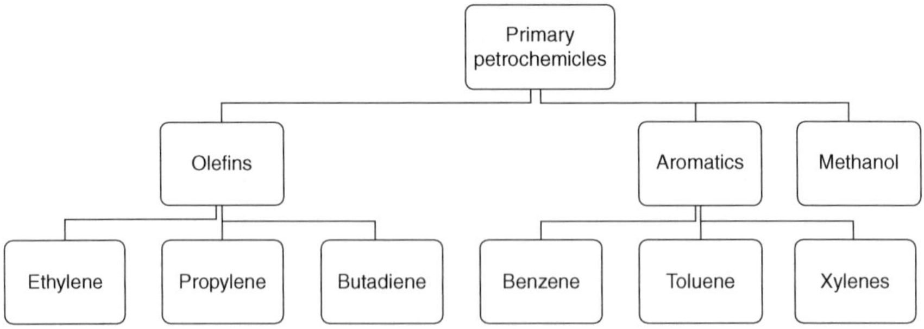

FIGURE 2.2 The Petrochemicals Family.

TABLE 2.1
World Consumption of Primary Petrochemicals by Region and Type: 2022

World Consumption of Primary Petrochemicals by Region: 2022

Region	Consumption (%)
Northeast Asia	50
North America	15
Western Europe	9
Middle East	9
Southeast Asia	7
Indian Subcontinent	4
Others	6

World Consumption of Primary Petrochemicals by Type: 2022

Petrochemicals by Type	Consumption (%)
Ethylene	32
Propylene	22
Methanol	16
Xylene	13
Benzene	10
Toluene	5
Butadiene	2

Source: www.spglobal.com/commodityinsights/en/ci/products/petrochemical-industry-chemical-economics-handbook.html

In 2022, Asia accounted for 60 percent of global naphtha consumption as a petrochemical feedstock. The region is expected to outperform other regions throughout the forecast period from 2024 to 2035 [4]. Asian petrochemical manufacturers depend more on naphtha compared to liquefied petroleum gas (LPG) and ethane, as refineries can supply naphtha in larger volumes. Despite having high naphtha yields, the region still imports large amounts of naphtha and LPG from the Middle East and Europe.

The choice of feedstock often depends on regional availability. Each region has developed its own feedstock specifics based on domestic market characteristics, available technology, and environmental standards.

Petrochemical industries take feedstock and use it to produce preliminary petrochemical products such as olefins, aromatics, and methanol, which are used as inputs in the manufacture of other petrochemical products. The market scope of each has been analyzed here.

2.4.1 Ethylene

There are several uses for the adaptable chemical ethylene. It is an essential raw ingredient in manufacturing textiles, resins, polymers, and other industrial chemicals. The expansion of end-use sectors such as consumer products, packaging, buildings, and automobiles affects the demand for ethylene. Global economic conditions, feedstock pricing, and supply-demand balance affect ethylene prices. Numerous businesses use ethylene-based products to keep up with the increasing demand for ethylene worldwide; the petrochemical industry frequently invests in ethylene production facilities and expands its capacity. Technological developments, market trends, and regulatory frameworks are observed to impact investment decisions.

Ethylene currently holds the largest share of the petrochemical market because of its versatility. It is one of the most important petrochemicals to produce polyethylene, ethylene oxide, and other derivatives, which are also used in packaging and the auto industry. The global production capacity of ethylene was 223.86 million metric tons in 2022, influenced by the fluctuations in crude oil prices (IMARC Group[5]) [5]. The market for ethylene is anticipated to grow at a Compound Annual Growth rate (CAGR) of 4.7 percent from 2022 to 2030 (Grand View Research[6]) [6, 7].

China is set to add as much production capacity for ethylene and propylene—as presently exists in Europe, Japan, and Korea combined. The demand for ethylene has increased mainly because of the fast growth of various industries such as construction, packaging, and transportation in emerging economies such as India, Brazil, Vietnam, and Thailand. The demand for ethylene fluctuates with growth in these sectors or shifts in consumer preferences.

On a global scale, North America ranks second in ethylene production capacity, following the Asia Pacific region. According to the Department of Energy in the United States, 95 percent of the ethylene production capacity in the United States is concentrated in the Gulf Coast region, specifically Texas or Louisiana. The United States is anticipated to raise its global market share for ethylene to 22 percent by the end of the year 2025 [8]. Due to the plentiful gas supply, the United States has an advantage in accessing low-cost ethane along with the Middle East.

2.4.2 Propylene

Propylene is extensively used in the automotive industry as the preferred choice for alloys, making up a significant portion of vehicle plastics. It is also a key component in manufacturing consumer goods, including housewares, bottle caps, toys, and luggage. The rising demand for polypropylene is also expected to drive the demand for propylene in Asia Pacific and the Middle East.

The propylene market is expected to be the second fastest-growing segment, driven by increasing demand for plastic manufacturing. According to the GVR report, propylene held over 20 percent share in the global industry in 2021. The market for propylene is anticipated to grow at a CAGR of 4.4 percent from 2022 to 2030. Propylene, a key feedstock for polypropylene production, can be produced by dehydrogenating propane. Polypropylene is widely used in packaging, automotive parts, and textiles. In 2021, Korea was the largest exporter of propylene, accounting for a share of 20.36 percent, trailed by the Netherlands, Taipei, Japan, and the United States, while the largest importer was China with 30.7 percent share, followed by Germany, Netherlands, and Belgium (GVR) [6].

2.4.3 Methanol

Methanol is utilized as a raw material in the production of acetic acid and formaldehyde. The majority of produced formaldehyde is employed in construction, remodeling, automobile and furniture manufacturing, and by original equipment manufacturers. The development of formaldehyde resins is closely associated with construction and remodeling operations and, to some extent, the automotive industry. In addition, methanol is widely used to manufacture biodiesel, which produces relatively fewer air pollutants than other fuels. The demand for biodiesel is expected to increase due to government initiatives aimed at reducing the global carbon footprint, which will subsequently boost the methanol market in the future.

In 2023, the global methanol market was worth US$31.26 billion. It is predicted to witness a CAGR of 4.5 percent from 2024 to 2032 (Fortune Business Insights) [9]. China's substantial investments in methanol to olefin projects contribute to its significant role in driving the methanol market. Asia, including China, accounts for over 70 percent of the global methanol demand, giving it a price advantage. Methanol production capacities are projected to increase in future, with major expansion plans in Russia, Iran, China, and the United States [6].

2.4.4 Butadiene

Butane and butylene are used to produce butadiene, essential for synthetic rubber, and isobutylene, which is used for fuel additives and other chemicals. There is a significant increase in demand for butadiene derivatives, particularly in Asian countries, due to the growing end-use industries such as construction, automobiles, consumer goods, etc.

In 2023, the revenue share of butadiene was the second-largest, accounting for greater than 14 percent share. The strong economic growth in emerging economies, along with the growing automotive industry, is expected to boost the growth of the butadiene market. It is estimated that during the years 2022 to 2030, the butadiene market will grow at a CAGR of 4.1 percent [6].

2.4.5 Benzene

Benzene is an important chemical used to produce nylon fibers, which are then used in textiles and engineering plastics. It also has applications in rubber, lubricants, dyes, detergents, drugs, and explosives processing. Benzene is a key component of aromatic petrochemical products. The demand for benzene is expected to increase due to its growing use in end-use industries such as inks, rubber, adhesives, paints, varnishes, as well as in industries like insulation and packaging. The increasing demand for adhesives and plastics in Asia Pacific and Latin America will also contribute to boosting the demand for these products.

In 2021, Benzene accounted for over 11 percent share in the global petrochemicals industry with a market size of 55.8 million tons. It is projected to grow at a CAGR of 4.1 percent during 2022–2030. In 2021, South Korea was the top exporter of benzene, with India, Thailand, and the Netherlands. China led in imports with 30.5 percent in 2021, trailed by the United States, Belgium, and the Netherlands [6].

2.4.6 Xylene

Xylene has a wide range of applications in the healthcare sector as a solvent. It is also used in the textile industry for manufacturing polyester. The growth of the textile industry in the Asia Pacific region, including India, Bangladesh, Thailand, Malaysia, and Sri Lanka, will boost xylene consumption in the future.

In 2021, the share of xylene in the petrochemical industry was over 5 percent. It is expected to increase at a CAGR of 5.4 percent during 2022–2030. P xylene is the major xylene product used worldwide, accounting for over 90 percent of the global xylene trade. In 2021, the leading importer of xylene was China, which had a 64.4 percent share. The top exporters of the product were Korea and Japan (GVR) [6].

2.4.7 Toluene

Toluene is usually produced via a catalytic reformer or ethylene cracker during gasoline manufacturing. Toluene is often utilized in industry as a substitute for benzene. It is a solvent commonly used in varnishes, paints, thinners, correction fluid, glues, and nail polish removers. It is also used in printing and leather tanning processes.

Due to the increasing application of toluene in pharmaceuticals, pesticides, and paints, the demand for toluene is estimated to increase in the future. However, market growth may be hampered by growing government regulations on environmental pollution.

In 2021, toluene made up over 10 percent of the global petrochemical industry. The leading toluene exporters were Japan, Thailand, Taipei, and Korea. India was the largest importer, accounting for over 19.5 percent followed by Korea, the United States, and China (GVR) [6].

2.5 PETROCHEMICAL INDUSTRY: KEY GROWTH DRIVERS

Refinery products are key growth drivers of the petrochemical industry, as they provide essential feedstocks for producing a wide range of petrochemicals. The feedstocks come from refining crude oil and processing natural gas. Approximately 90 percent of all raw materials used as feedstock are derived from oil and gas, with the remaining 10 percent coming from coal and biomass.

Let us discuss how these refinery products contribute to the growth of petrochemical units.

1. **Essential Feedstocks for Production:** Naphtha is a highly versatile feedstock used in steam crackling to produce ethyne, propylene, and other olefins. These olefins are fundamental building blocks for various petrochemicals and plastics. In 2021, naphtha was the most widely used feedstock globally, followed by natural gas, coal, and ethane [2]. Ethane and naphtha are the primary feedstocks used to produce high-value chemicals (HVCs), while natural gas is the primary input for the global production of methanol and ammonia, except in China, where coal is predominantly used for both. NGLs are essential raw materials for steam cracking and other processes, resulting in a wide range of petrochemicals.
2. **Economies of Scale and Integration:** Refineries often integrate with petrochemical units to optimize the use of intermediate products. For instance, a refinery can direct naphtha and other intermediates to adjacent petrochemical plants, reducing transportation costs and improving efficiency.

   Large integrated complexes benefit from economies of scale, reducing feedstock's per-unit cost and enhancing petrochemical production's competitiveness.
3. **Development in Technology:** Technological advancements in catalysis crackling processes have improved the yield and efficiency of converting refinery products into valuable petrochemicals. Innovations in these technologies enable the production of higher-value chemicals from the same feedstock volume. Refineries are implementing cleaner production technologies to minimize environmental impact, benefiting petrochemical units with more sustainable and compliant feedstocks.
4. **Market and Economic Dynamics:** The availability and pricing of refinery products have a significant impact on the competitiveness of petrochemical units. Regions with abundant, low-cost feedstocks, such as North America with its shale gas boom, can produce petrochemicals at a very low cost, thus enhancing their global market position. Additionally, investment in refining and petrochemical infrastructure in regions like the Asia Pacific and the Middle East is driving growth by expanding their refining capacities to supply feedstocks to growing petrochemical industries.
5. **Innovation and Diversification:** The diverse range of refinery products allows petrochemical units to develop new products and applications. For instance, specialized chemicals and advanced polymers are continuously being developed to meet emerging market needs.

   Innovations in bio-based and recycled feedstocks complement traditional refinery products, helping petrochemical units diversify their feedstock base and meet sustainability targets.

2.6 GLOBAL PETROCHEMICAL MARKET: BY REGION

The petrochemicals market is segmented into five major regions: Asia Pacific, North America, Europe, Latin America, and the Middle East and Africa. The increasing demand across various industries and regions is the driving force behind the global petrochemical market. However, the market dynamics are complex and influenced by geopolitical factors, regional capacities, technological advancements, and regulatory landscapes.

In 2023, the global petrochemicals market was worth US$623.82 billion and is projected to grow at a CAGR of 4.2 percent from 2024 to 2032, according to Fortune Business Insights [9]. See Table 2.2.

According to Research Nester [10], the petrochemical market in the Asia Pacific region is expected to experience moderate progress due to rapid economic development in the area. The global petrochemical market in the North American region is likely to grow because of the increasing demand for plastic, fertilizer, synthetic rubber, and detergent.

Let us discuss the factors affecting demand for petrochemical products in different regions.

2.6.1 Asia Pacific

The main player in the petrochemical market is Asia Pacific, which offers numerous business opportunities due to rising plastic packaging demand. The increasing need for packaging applications is projected to surge the demand for petrochemical products from 2024 to 2032. In 2023, the market size in the Asia Pacific region was recorded at US$325.44 billion (Fortune Business Insights) [9]. The robust growth of the Asia Pacific plastic packaging industry is anticipated to positively impact the petrochemicals market. China is exporting cheaper plastics to other regions,

TABLE 2.2
Global Petrochemical Market Share (in %); Segmentation by Region (Projections for 2031)

Region	Global Petrochemical Market Share (%) by Region 2031
Asia Pacific	29
North America	33
Europe	22
Latin America	9
Middle East and Africa	7

Source: Research Nester www.researchnester.com/reports/petrochemical-market/4125

capturing the market share from South Korea and Japan. To meet the increasing demand, companies in this area increasingly use non-oil feedstock and NGLs. They are also focusing on cost-effective strategies to boost sales.

Favorable government regulation, rapid industrialization, urbanization, and low costs are expected to drive petrochemical demand in Asia Pacific countries, including China, India, Thailand, and Indonesia. Increasing spending on construction activities in these regions for developing residential and commercial buildings is expected to drive petrochemical demand.

According to GVR, the petrochemicals market in the Asia Pacific region held over 51.0 percent share in 2023, mainly due to the thriving chemicals industry and growing polymer consumption (GVR) [6].

2.6.2 Middle East

The Middle East is a major hub of petrochemical products. Middle Eastern nations are the primary exporters of crude oil and are accountable for nearly 40 percent of crude oil production globally. Rapid urbanization, population growth, and positive developments in these countries (UAE, Qatar, and Bahrain) enhance the demand for petrochemical products. With the oil plants shut down in Europe and the decline in Russian oil output, the Middle East is shifting its emphasis towards petrochemical ventures instead of refining. Furthermore, Gulf nations are efficiently meeting rising demand by capitalizing on raw materials, extensive energy reservoirs, and cost advantages compared to developed countries.

2.6.3 North America

Over the past decade, technological advancements in North America have significantly boosted the region's petroleum sector. The abundance of low-cost shale gas has been particularly advantageous for the North American market. In 2023, the United States held around 40 percent of global ethane-based petrochemical production capacity (IEA) [1]. The country is one of the top ten producers of petrochemicals demanded by the automotive, construction, paints, and coating industry. It is also a significant exporter of petrochemical products, taking advantage of its low production costs.

2.6.4 Europe

Petrochemicals accounted for more than one-fourth of the entire European chemical market in 2023. Europe is likely to experience a CAGR of 6.8 percent from 2024 to 2030 (GVR) [6]. The European market is projected to grow steadily but is currently experiencing structural changes due to increasing energy and feedstock prices as well as growing pressure from other countries.

In Europe, this sector is heavily regulated due to concerns about health, the environment, climate change, safety, and energy. High energy costs and strict regulations present challenges to the region's competitiveness. Consequently, Europe is expected to see a decrease in its share of the global market from 2024 to 2032. To maintain its position in the market, Europe's petrochemical industry is prioritizing sustainability and innovation by investing in green and innovative technology.

2.6.5 South America

According to the International Monetary Fund (IMF), economic growth in South America slowed down in 2023 compared to 2022. Furthermore, Brazil and Argentina are expected to experience an economic slowdown in the coming years, which may impact the growth of this sector. The market in this region is likely to grow below average from 2024 to 2032 owing to poorly developed industries (Fortune Business Insights) [9]. However, the increasing demand for products like plastics, dyes, and thinners from the automotive and construction sectors is projected to drive the growth of the petrochemical sector in the region.

2.7 PETROCHEMICALS AND STANDARD OF LIVING

Petrochemicals significantly improve living standards by making essential goods and services more affordable, accessible, and effective. The consumption of petrochemicals enhances the quality of life by contributing to healthcare, housing, transportation, food security, consumer goods, energy, and technology.

1. **Healthcare and Medicine**
   Petrochemicals play a crucial role in the healthcare industry by producing many pharmaceutical products and medicines. They are used to manufacture active pharmaceutical ingredients (APIs) and various solvents and excipients that are crucial in drug formulation.

   Many medical devices and equipment, such as syringes, IV bags, tubing, and prosthetics, are made from petrochemical-derived plastics and polymers. These materials are critical for ensuring sterility and functionality in medical settings.

   Personal protective equipment (PPE) like masks, gloves, and gowns, vital for infection control, especially highlighted during the COVID-19 pandemic, are produced using petrochemical products.
2. **Food and Agriculture**
   Petrochemicals are integral to food supply chain and agriculture. Plastics derived from petrochemicals are widely used in food packaging, which helps preserve food, reduce spoilage, and extend shelf life. This contributes to food safety and reduces waste. Petrochemical products manufacture machinery and equipment involved in food processing and storage. Polyethylene and polystyrene are commonly used in food packaging.

   Packaging is the largest consumer of plastic globally. The industry's growth is tied to increasing global consumption and urbanization. The packaging sector is projected to dominate the petrochemicals market in 2023 (Fortune Business Insights [9]. This segment includes both consumer packaging and business-to-business transactions. E-commerce has revolutionized the packaging industry, as companies look for tailored solutions to protect products during transportation and enhance the consumer experience.

Petrochemicals are key ingredients in producing fertilizers and pesticides, which enhance crop yields and improve food security.

3. **Housing and Infrastructure**

   Petrochemicals are used to produce a variety of construction materials, including insulation, pipes, adhesives, and paints. These materials are crucial for building safe, durable, cost-effective, and energy-efficient homes and buildings.

   Petrochemical products are used in constructing roads, bridges, and other infrastructure projects, essential for economic development and improving living standards.

4. **Consumer Goods**

   Many consumer goods, from electronics appliances to clothing and toys, are made using plastics and polymers derived from petrochemicals. These products enhance convenience, comfort, and lifestyle.

   Household Products like cleaning agents and personal care items are often made using petrochemicals, contributing to hygiene, timesaving consumer goods, and overall well-being.

5. **Transportation**

   Petrochemicals contribute to transportation through fuels and materials. While the focus is often on renewable energy, traditional fuels derived from petrochemicals (gasoline, diesel, and jet fuel) still play an important role in transportation. They enable the movement of people and goods, which is fundamental to economic activity and access to services.

   Various automotive parts are manufactured using lightweight, durable plastics, which help to reduce vehicle weight and improve fuel efficiency, safety, and comfort in modern transportation.

6. **Energy and Utilities**

   Petrochemicals play an important role in generating energy, both directly as fuels and indirectly in manufacturing components for renewable energy systems like wind turbines and solar panels.

   Petrochemical products are used in producing batteries and other energy storage solutions, which are critical for the reliability and efficiency of modern energy systems.

7. **Communications and Technology**

   Many electronic devices, including smartphones, computers, and televisions, rely on petrochemical-derived components such as plastics, circuit boards, and semiconductors.

   Petrochemical products are essential in manufacturing cables, connectors, and other components necessary for telecommunications infrastructure. These products enhance functionality and affordability.

8. **Sustainability and Recycling**
   Advances in petrochemical technology have made it possible to recycle plastics and other materials more efficiently, reducing waste and environmental impact.
   Bio-based Alternatives: Research and development in the petrochemical industry are leading to the creation of bio-based plastics and chemicals, which can provide more sustainable alternatives to traditional petrochemical products.

2.8 KEY PLAYERS IN THE GLOBAL PETROCHEMICALS INDUSTRY

The petrochemical industry is highly competitive. Several key players in the global petrochemical market contribute significantly to the production and distribution of petrochemical products worldwide. The key market players in the global petrochemicals industry organized by the country are as follows:

1. **United States**
   a. ExxonMobil Chemical (www.exxonmobilchemical.com/en/)
   b. Dow Chemical Company (www.dow.com/en-us)
   c. Chevron Phillips Chemical Company (www.cpchem.com/)
   d. DuPont de Nemours, Inc. (www.dupont.com/)
2. **Saudi Arabia**
   a. Saudi Basic Industries Corporation (SABIC) (www.sabic.com/en)
   b. The Saudi Aramco Total Refining and Petrochemical Company (SATORP) (www.satorp.com)
3. **China**
   a. Sinopec (China Petroleum and Chemical Corporation) (www.sinopecgroup.com/group/en/)
   b. China National Petroleum Corporation (CNPC) (www.cnpc.com.cn/en/)
4. **South Korea**
   a. LG Chem (www.lgchem.com)
   b. SK Innovation (www.skinnovation.com/)
   c. Lotte Chemical (www.lottechem.com)
   d. Hanwha Chemical (https://hcc.hanwha.co.kr/en/)
   e. S-Oil Corporation (www.s-oil.com/en/)
5. **Germany**
   a. BASF SE (https://basf.com)
   b. Covestro AG (www.covestro.com/en)
   c. Evonik Industries AG (https://corporate.evonik.com/en)
6. **Japan**
   a. Mitsubishi Chemical Corporation (www.mcgc.com/english/)
   b. Sumitomo Chemical Company, Limited (www.sumitomo-chem.co.jp/english/)

   c. Mitsui Chemicals, Inc. (https://jp.mitsuichemicals.com/en)
   d. Idemitsu Kosan Co., Ltd. (www.idemitsu.com/en)
7. **India**
   a. Reliance Industries Limited (Petrochemicals Division) (www.ril.com/businesses/petrochemicals)
   b. Indian Oil Corporation Limited (Petrochemicals Division) (https://iocl.com/pages/Petrochemicals-Overview)
   c. Haldia Petrochemicals Limited (www.haldiapetrochemicals.com/)
   d. ONGC Petro additions Limited (www.opalindia.in/)
8. **United Arab Emirates**
   a. Abu Dhabi National Oil Company (ADNOC) (www.adnoc.ae/)
   b. Borouge (Joint venture between ADNOC and Borealis) (www.borouge.com/en)
9. **Netherlands**
   a. Royal Dutch Shell (Chemicals Division) (www.shell.com/business-customers/chemicals.html)
   b. LyondellBasell Industries (www.lyondellbasell.com/)
10. **Russia**
   a. Sibur (www.sibur.com/en/)
   b. Lukoil (Petrochemicals Division) (www.lukoil.com/Products/business/petrochemicalproduction)
   c. Gazprom (Petrochemicals Division) (https://salavat-neftekhim.gazprom.com/about/working/chemistry/)

As shown in Table 2.3, BASF, based in Germany, has maintained its position as the largest chemical company in the world in terms of annual revenue for several consecutive years. The company's diverse portfolio has resulted in impressive financial performance, with revenue exceeding US$94.83 billion in 2023. Dow Chemical and LyondellBasell, both headquartered in the United States, have also established themselves as leading global chemical companies, consistently ranking among the top five (Statista Research Department) [11].

2.9 FUTURE OF PETROCHEMICALS

Petrochemicals play a crucial role in various industrial processes and serve as the foundation of an economy. The production of petrochemicals is also a significant consumer of energy and resources, making it an important sector for sustainable development. With the growing environmental concerns, the petrochemical industries increasingly focus on safe, sustainable, and eco-friendly production, which drives the industry's development of new technologies and innovations. To tackle the challenges of an uncertain and volatile market, petrochemical companies should prioritize operations, maintenance, supply chain, procurement, and talent management.

TABLE 2.3
Global Leading Chemical Companies Based on Revenue: 2023 Ranking Here

Global Leading Chemical Companies	Revenue (in Billion US$)
BASF, Germany	94.83
DOW, US	56.9
LyondellBasell Industries, US	50.45
LG Chem, South Korea	39.27
Nutrien, Canada	37.87
Mitsubisi Chemical Holding	33.55
Linde, Ireland	33.36
Air Liquide, France	32.51
Umicore, Belgium	31.52
Hengli Petrochemical, China	31.51

Source: www.statista.com/markets/410/topic/445/chemical-industry/#statistic4

The supply and demand dynamics in the chemical sector are changing global competition. The Middle East, particularly Iran and Saudi Arabia, is seen as another low-cost region for various petrochemicals, with numerous global projects planned. Europe and China together account for approximately one-fourth of the global capacity for high-value, naphtha-based chemicals.

2.9.1 A Cut in Fuel Demand Pushes Petrochemicals Industry

According to a survey, the need for gasoline is projected to decrease by 15 percent from 2018 to 2035, and the demand for diesel is expected to decrease by 6 percent. Conversely, the demand for petrochemicals is expected to grow by 3 to 6 times, largely due to ongoing technological advancements, the increased use of autonomous and shared vehicles, the transition to renewable energy sources, and expanding middle class in developing nations [12]. To improve investment profitability, companies can expand into petrochemicals. To offset the predicted decline in US fuel consumption and shrinking export markets, refiners may reduce crude capacity or diversify into higher-value petrochemicals. The latter option will allow them to preserve crude capacity and achieve better margins than producing fuels.

2.9.2 Rising Demand for Electronic Products and Household Goods

Petrochemicals play a crucial role as raw materials in electronics and other household goods, and the rising demand for electronic products will further boost the petrochemicals market. The push towards electric vehicles (EVs) also boosts the demand for specialized plastics and composites. Plastics are widely used to produce bumpers, interior panels, dashboards, and non-metallic parts used in automotive production. These materials help reduce total vehicle weight, increase fuel efficiency, and meet strict emission standards. The demand for plastics has surpassed steel, aluminum, and cement. The growth of EVs in China and its emissions target is likely to benefit exporters of petrochemicals to China.

The demand for petrochemicals in critical end markets like packaging and construction will continue growing due to economic development in China, India, and Southeast Asia. According to the Future of Petrochemical report, developed economies currently use up to 20 times more plastic and about 10 times more fertilizer per capita than India, Indonesia, and other developing countries [1]. This highlights the significant potential for global growth. As life expectancy and standard of living increase, people are likely to consume more of these products.

2.10 BUSINESS CHALLENGES AND PROSPECTS

2.10.1 Environmental Issues and Regulations

The petrochemical industry is under heightened scrutiny due to environmental concerns. Strict regulations and targets for reducing carbon emissions will require substantial investments in technologies, sustainable practices, and waste management, which will lead to an increase in cost of production. The industry's transition towards sustainability and innovation is expected to significantly influence the future demand and supply of petrochemical products.

The petrochemical industry faces various challenges, including climate impact, water pollution, and air quality issues. Despite these challenges, petrochemical products offer advantages, such as their increasing use in advanced technologies. Rising concerns about the health and environment and increasing demand for bio-based and renewable alternatives may impact the growth of the petrochemical market.

2.10.2 Impact of War Between Russia and Ukraine

The war between Russia and Ukraine has significantly impacted the petrochemical industry by creating uncertainties and disrupting the supply chain of petrochemicals. The conflict led to an increase in the price of natural gas, a crucial feedstock for the petrochemicals industry. As a result, higher costs reduced profit margins, making it more challenging for companies to remain competitive. In 2022, the shortfall in global oil supplies from Russia drove oil prices to over $120 a barrel. However, as the global economy slowed down, prices sharply reduced. The petrochemicals and polymers industry experienced severe impacts from the war, resulting in supply-chain issues and lower overall government stimulus (ICIS 2022) [13].

2.10.3 Economic Uncertainty

Economic uncertainty and weak operating rates may limit petrochemical recovery in 2024. Due to severe inventory destocking, demand for most petrochemicals remained weaker than expected in 2023. In the upcoming year, rising inflation, increasing interest rates, and escalating geopolitical tensions pose significant challenges and risks to economic growth (S&P Global). An increasing mismatch between demand and supply will also affect utilization and product profit margin.

2.10.4 Use of ML and AI in Manufacturing and Data Mining

Machine learning (ML) techniques mimic human thinking, detect patterns, and mine data faster than humans do. These technologies are utilized to help predict process outcomes, develop hypotheses, and identify gaps in scientific literature. There is a trend in the chemical sector that incorporates Artificial Intelligence (AI), the Internet of Things (IoT), and ML. These advanced technologies improve efficiency, productivity, and reduce energy loss and risk.

2.11 CONCLUSION

The petrochemical industry has had a significant impact on global economic development. It plays a major role in generating employment and advancing various sectors. The industry's capacity to innovate and adjust to changing market demands underscores its crucial role in economic growth. The industries have also recognized the advantages of utilizing data and analytics to enhance production efficiency, product quality, output, and reliability (Mckinsey) [14]. However, this progress is accompanied by challenges, such as environmental concerns and the need for sustainable practices. In order to ensure that the petrochemical industry makes a positive contribution to a sustainable and prosperous future, it is crucial for policymakers, industry leaders, and communities to work together to balance economic benefits with environmental stewardship. The industry must prepare for a more challenging, competitive environment. It is essential to maximize potential and achieve excellence in operations, maintenance, supply chain, procurement, capital spending, and talent management to promote sector growth.

NOTES

1. The Future of Petrochemicals: Towards more sustainable plastics and fertilizers, OECD/IEA 2018.
2. Petrochemical Industry Overview, Chemical Economics Handbook.
3. www.bloomberg.com/news/articles/2023-07-09/a-glut-of-made-in-china-plastic-will-dent-oil-s-growth-machine.
4. Olefins, aromatics, and methanol are variety of chemical products generally known as primary petrochemicals.
5. International Market Analysis Research and Consulting (IMARC) Group.
6. Petrochemical Industry Outlook Data Book, 2023–2030.

REFERENCES

1. OECD/IEA. (2018). The future of petrochemicals: Towards more sustainable plastics and fertilizers. International Energy Agency. www.iea.org/reports/the-future-of-petrochemicals
2. S&P Global. (n.d.). *Chemical Economics Handbook*. www.spglobal.com/commodityinsights/en/ci/products/petrochemical-feedstocks-chemical-economics-handbook.html
3. Low, E., & Crowley, K. (2023, July 9). A glut of made-in-China plastic will dent oil's growth machine. *Bloomberg*. www.bloomberg.com/news/articles/2023-07-09/a-glut-of-made-in-china-plastic-will-dent-oil-s-growth-machine
4. ChemAnalyst. (n.d.). *Naphtha market analysis*. www.chemanalyst.com/industry-report/naphtha-market-637
5. IMARC Group. (n.d.). *Chemicals market reports*. www.imarcgroup.com/categories/chemicals-market-reports
6. Grand View Research. (n.d.). *Petrochemicals Industry Data Book – Ethylene, Propylene, Butadiene, Benzene, Xylene, Toluene and Methanol Market Size, Share, Trends Analysis, and Segment Forecasts, 2023–2030*. www.grandviewresearch.com/sector-report/petrochemicals-industry-data-book
7. Grand View Research. (n.d.). *Global plastics market*. www.grandviewresearch.com/industry-analysis/global-plastics-market
8. U.S. Department of Energy. (2022). *U.S. ethane: Market issues and opportunities*. www.energy.gov/sites/default/files/2023-06/U.S.Ethane-Market-Issues-and-Opportunities.pdf
9. Fortune Business Insights. (n.d.). *Petrochemicals market size, share & industry analysis*. www.fortunebusinessinsights.com/petrochemicals-market-102363
10. Research Nester. (n.d.). *Petrochemical market*. www.researchnester.com/reports/petrochemical-market/4125
11. Statista Research Department. (2023, May 21). Global list of leading chemical companies based on revenue. *Statista*. www.statista.com/topics/2026/top-chemical-companies/#topicOverview
12. Precedence Research. (n.d.). *Petrochemical market*. www.precedenceresearch.com/petrochemical-market
13. ICIS. (2022, March). Ukraine-Russia: How the crisis could reshape petrochemicals demand. *Asian Chemical Connections*. www.icis.com/asian-chemical-connections/2022/03/ukraine-russia-how-the-crisis-could-reshape-petrochemicals-demand/
14. McKinsey & Company. (n.d.). Excellence in petrochemicals: What it will take to win. www.mckinsey.com/industries/chemicals/our-insights/excellence-in-petrochemicals-what-it-will-take-to-win#/

Disclaimer: All brand logos, trademarks, company names, company photographs, and company data mentioned in this chapter are the intellectual property of their respective owners. These are used for informational and educational purposes only and are not intended to imply any endorsement or affiliation with the brands/companies. The inclusion of this information is for illustrative purposes to support the analysis and discussion in this academic work. No commercial use or exploitation of the logos, trademarks, company names, company photographs, and company data is intended or should be inferred.

3 Petrochemical Complexes

3.1 INTRODUCTION

Petrochemical complexes, or integrated petrochemical complexes or petrochemical hubs, are industrial facilities that incorporate multiple petrochemical production units within a single site or location. These complexes are strategically designed to optimize economies of scale, enhance operational efficiency, and facilitate synergy among various production processes. Some of the critical units typically found in petrochemical complexes are:

1. **Refinery Units:** Petrochemical complexes often include refining units to process crude oil into intermediate feedstocks such as naphtha, ethane, propane, and butane. It may consist of crude distillation units, catalytic crackers, reformers, and hydro-treaters to produce feedstocks suitable for petrochemical production.
2. **Petrochemical Plants:** Various petrochemical production units are integrated into the complex to convert feedstocks into primary petrochemicals and intermediate chemicals. Some of the common petrochemical plants include steam crackers for ethylene and propylene production, aromatics plants for benzene, toluene, and xylene (BTX), and other units for producing olefins, polyolefins, aromatics, and specialty chemicals.
3. **Polymerization Units**: Polymerization units in the petrochemical complexes use the chemical processes which convert basic petrochemicals and intermediates into polymer resins. The units produce the products like polyethylene (PE), polypropylene (PP), polyvinyl chloride (PVC), and polystyrene (PS). It produces various grades of polymers having specific properties. The characteristics of various polymer resins are tailored to customer needs and their applications.
4. **Intermediate Petrochemicals Chemical Units:** The intermediate chemicals like ethylene oxide (EO), ethylene glycol, styrene, methanol, and acrylonitrile are produced with the help of additional chemical processing units. These intermediate chemicals are used as a raw material for downstream industries. They are used in plastics, textiles, automotive, and pharmaceutical industries.

DOI: 10.1201/9781032661933-3

5. **Utilities and Infrastructure:** Petrochemical complexes require extensive infrastructure to support production operations, including utilities such as steam boilers, power plants for electricity, water, cooling towers, water treatment facilities, and compressed air. Infrastructure includes storage tanks, loading/unloading facilities, pipelines, and transportation networks for raw materials, intermediates, and finished products. Shared utility infrastructure allows for economies of scale, cost savings, and operational flexibility across multiple petrochemical units within the complex.
6. **Logistics and Supply Chain Facilities:** Petrochemical complexes may feature logistics hubs, warehouses, and distribution centers to manage inbound and outbound logistics, inventory storage, and transportation of raw materials and products. To ensure timely delivery of feedstock and finished products to end customers the petrochemical complexes need efficient logistics and supply chain facilities.
   Petrochemical complexes facilitate integration along the value chain by providing feedstocks to downstream manufacturing industries and receiving by-products or co-products from adjacent facilities. Integration with downstream industries such as plastics, chemicals, textiles, and automotive enables the efficient utilization of petrochemical products and promotes industrial symbiosis.
7. **Research and Development Centers:** Some petrochemical complexes incorporate R&D centers or laboratories dedicated to innovation, process optimization, product development, and technology advancement. These facilities are crucial in driving continuous improvement, developing new products, and enhancing operational excellence within the complex.
8. **Support Services and Amenities:** Petrochemical complexes provide support services and amenities for employees, including administrative offices, training facilities, medical clinics, cafeterias, and recreational areas. Safety and security measures are implemented to protect personnel, assets, and the surrounding environment from operational risks and hazards.

Petrochemical complexes serve as critical hubs for petrochemical production, leveraging integrated infrastructure, synergistic processes, and centralized operations to optimize efficiency, competitiveness, and sustainability in the global petrochemical industry.

Figures 3.1 and 3.2 are illustrative pictures of the petrochemical complexes.

3.1.1 Characteristics of Petrochemical Complexes

The key characteristics of petrochemical complexes are as follows:

1. **Infrastructure and Location:** Petrochemical complexes are typically located near major sources of feedstock supply, such as oil refineries or natural gas processing plants, to minimize transportation costs and logistics.

FIGURE 3.1 Petrochemical Complex.

FIGURE 3.2 Petrochemical Complex.

Other feedstocks for petrochemicals include natural gas liquids, naphtha, and other hydrocarbons. Integration with upstream facilities allows for the direct supply of feedstocks to petrochemical units, minimizing transportation costs and logistical complexities. They may also be situated near key markets and transportation networks like roads and ports to facilitate the distribution of products to customers.

The infrastructure within petrochemical complexes includes utilities such as power generation, steam, water, and wastewater treatment facilities, as well as storage tanks, pipelines, and transportation terminals.

2. **Design for Integration and Synergies:** The petrochemical complexes are planned such that various production units are put up together to optimize the resource utilization and maximize the synergies. Multiple plants are set up in close proximity for economies of scale and to optimize energy usage. Infrastructure and utilities are used on a sharing basis to minimize the environmental impact. Integrating the facilities helps in by-product utilization, flexibility in feedstock, and recovery of heat waste. Thus the integration of facilities enhances the overall efficiency and competitiveness.
3. **Diversified Production Units:** As discussed in the earlier paragraphs the petrochemical complexes have diversified units which produce basic petrochemicals (like ethylene, propylene, benzene), intermediate chemicals (methanol, styrene, EO, etc.), and polymer products. The production units planned in these complexes are steam crackers, aromatic units, polymerization facilities, olefin plants, etc. It also can have downstream processing units to offer finished products to end customers. The production units are normally connected with feedstock storage tanks and utility systems through pipelines for material transfer to ensure seamless operations. Diversified units located on the same premises lead to productive utilization of utilities, infrastructure, and resources.
4. **Value-Added Facilities:** The petrochemical complexes also have R&D facilities and product testing laboratories. They also offer technical support services to manufacturing department and have facilities to organize training programs for various stakeholders.

   The services like technical support and training help in fostering innovation, collaboration, and technology transfer within the ecosystem of the complex. It also drives continuous improvement and competitiveness.
5. **Development of Cluster:** The industrial cluster gets developed around petrochemical complexes. Clusters may include equipment manufacturers, catalysts, chemicals, packaging materials, logistics providers, and research institutions. Related industries and service providers establish their operations in the proximity of petrochemical complexes to create industrial hubs/clusters. The development of cluster promotes synergistic collaboration among companies in petrochemical value chain. It also offers opportunities for vertical and horizontal integration and supply chain efficiencies.
6. **Environmental and Regulatory Compliance:** Environmental and regulatory compliances are with respect to emission controls, waste management, and safety aspects of occupational hazards. Petrochemical complexes use advanced pollution control technologies and emergency response protocols to comply with these requirements. The companies develop (i) environmental management systems (EMS), (ii) monitoring programs, and (iii) implement emergency response plans to mitigate risks

and ensure environment and regulatory compliance. To promote social responsibility sustainability initiatives like energy efficiency and waste minimization are encouraged.

7. **Investment in Petrochemical Complexes:** These complexes require huge infrastructure in terms of land and other related facilities with the latest technology with a future plan to expand operations.
   Governments support such developments of petrochemical complexes through incentives and various policy measures related to infrastructure investment. Industry stakeholders and investors collaborate to develop petrochemical complexes. The feasibility and viability of such projects require long-term planning, market analysis, and risk assessment and mitigation.
8. **Economic Benefits:** The country or region benefits from investment in Petrochemical complexes as it contributes significantly to regional economic development by attracting investments. It also creates employment opportunities in the region, and stimulates the downstream industries.

Petrochemical complexes are strategic assets that play a pivotal role in the global petrochemical industry, fostering innovation, collaboration, and sustainable development. By leveraging integration, synergies, and value-added services, these complexes contribute to economic growth, industrial competitiveness, and value creation across the petrochemical value chain.

Examples of some of the notable petrochemical complexes include:

1. Jamnagar Petrochemical Complex in India [1]
2. Jurong Island Petrochemical Complex in Singapore [2]
3. Texas Gulf Coast Petrochemical Complex in the United States [3, 4]
4. Jubail Industrial City Petrochemical Complex in Saudi Arabia [5, 6]
5. Nanjing Petrochemical Complex in China [7, 8]

Petrochemical complexes play a vital role in the global industry by providing integrated and efficient manufacturing hubs for producing a wide range of petrochemical products essential for various industrial sectors.

3.2 MAJOR PETROCHEMICAL CAPACITIES ACROSS THE GLOBE

The total worldwide aggregate capacity of petrochemicals was about 2,600.06 MMTPA in the year 2023 [9]. It is important to know that the capacities of petrochemical complexes can vary over time due to expansions, upgrades, or changes in production configurations. Therefore the capacities mentioned in the following paragraphs are aggregate in nature and it is for the purpose of broader understanding.

Brief details, aggregate capacities, and range of products produced by some major petrochemical complexes listed in the above paragraph are given below.

Jamnagar Petrochemical Complex, India: This complex is operated by Reliance Industries Limited; the Jamnagar complex is one of the largest refining and petrochemical complexes in the world. It has significant capacities for producing petrochemicals such as ethylene, propylene, PE, PP, and other downstream products. The combined petrochemical capacity of this complex is over 12.56 million metric tons per year (MMTPA)[10].

Jurong Island Petrochemical Complex, Singapore: Jurong Island in Singapore is a major petrochemical hub in Asia, hosting various petrochemical companies and facilities. It houses many petrochemical companies, three major players who have a presence in this island are ExxonMobil, Shell, and BASF.

The other heavyweights who are present in this petrochemicals hub are DuPont, Huntsman, Perstorp, Dainippon Ink & Chemicals, and Sumitomo Chemical. Shell has its 750,000 tpa mono ethylene glycol (MEG) plant on Jurong Island. It is one of the world's largest MEG facilities. It is one of the key components of the Shell Eastern Petrochemical Complex (SEPC).

Exxon Mobil has established 1 MTPA ethylene steam cracker, this project uses ExxonMobil's proprietary technologies. The company Lanxess which is specializing in the development, manufacturing, and marketing of plastics, rubber, intermediates, and specialty chemicals has built a butyl rubber facility of 0.1 MTPA in Singapore. Lanxess plant serves the rising demand for tires brought on by increasing use of cars, particularly in India and China.

This complex has capacities for producing a wide range of petrochemicals, which includes ethylene, propylene, benzene, toluene, and xylene. It also produces olefins, aromatics, and specialty chemicals. The combined aggregate capacity of petrochemical plants on Jurong Island exceeds 30 MMTPA.

Gulf Coast Petrochemical Complex, Texas, United States: This hub has a number of petrochemical complexes operated by companies like ExxonMobil, Dow Chemical, Chevron Phillips Chemical, and Lyondell Basell. It has significant production capacities to produce ethylene, propylene, PE, PP, styrene, and other petrochemicals.

ExxonMobil and SABIC have jointly set up the manufacturing facility in this complex with equal share in ownership. ExxonMobil serves as the site operator. This facility is set up to meet the demand for petrochemical products which are required for the manufacturing of packaging, construction materials, clothing, coolants, and other production types. It has a 1.8 MMTPA ethane steam cracker, two PE units with a combined capacity for 1.3 MMTPA, and a monoethylene glycol unit with a capacity for 1.1 MMTPA. The aggregate total petrochemical capacity in the Texas Gulf Coast region is estimated to be over 4.99 MMTPA [11, 12].

Jubail Industrial City Petrochemical Complex, Saudi Arabia: It is located in Jubail Industrial City, Saudia Arabia. This city is located in the Eastern Part of Saudi Arabia. Major industries operating in this city are SABIC, Saudi Aramco, Al-Jubail Petrochemical Company, Arabian Petrochemical Company (Petrokemya), and Jubail United Petrochemical Company. Total Refining and Petrochemical

Company (Satorp) refinery and Sadara Chemical Company petrochemicals complex are also located in this city. It is one of the largest integrated petrochemical complexes in the Middle East. The products offered by this complex are ethylene, propylene, PE, PP, methanol, and fertilizers. The aggregate petrochemical capacity of this complex is in excess of MMTPA.

Nanjing Petrochemical Complex, China: Nanjing Petrochemical Complex is located in Jiangsu Province of China. It houses several petrochemical complexes operated by many players in the industry. The major players are Sinopec, Yangzi Petrochemical, and BASF. It produces various petrochemicals like ethylene, propylene, PE, PP, ethylene vinyl acetate (EVA), EO, etc. The aggregate total petrochemical capacity of this complex is estimated to be more than 10 MMTPA. This petrochemical complex serves both domestic and international markets. Further, the capacities of petrochemical complexes worldwide vary significantly depending on factors such as feedstock availability, market demand, investment trends, and technological advancements.

BASF-YPC Co Ltd (BASF-YPC), which is operating in this complex is a chemical company that distributes specialty chemicals. The company offers products such as glacial acrylic acid, n-butyl acrylate, dimethyl amino ethyl acrylate, methyl acrylate, toluene, and benzene and polymer grade propylene. It also provides polymer grade ethylene, mixed xylenes, butadiene, i-butanol, n-butanol, propionaldehyde, isobutyraldehyde, n-butyraldehyde, propionic acid, formic acid, triethylene glycol, and PS. These products are used in various applications including rubbers, superabsorbent polymers, detergents, dispersions for adhesives, coatings, and paints, paper sizing and non-woven plastics, and PAN-fibers. The company markets its products to chemicals and polymers industries.

3.2.1 Key Petrochemical Products: Production Capacities

Global petrochemical capacities are significant, reflecting the industry's importance in supplying essential materials for various sectors.

Ethylene

Ethylene is one of the primary building blocks in the petrochemical industry, and it is used in the production of various plastics, fibers, and chemicals. Global ethylene production capacity is estimated to exceed 200 million metric tons per year (MMTPA) [13]. Major ethylene producers include the United States, the Middle East (notably Saudi Arabia and Qatar), China, and Europe.

Propylene

Propylene is another key petrochemical used in manufacturing plastics, synthetic fibers, and chemicals. Global propylene production capacity is estimated to be around 100–110 MMTPA [14]. Significant propylene capacity is concentrated in regions with extensive refining and petrochemical industries, such as the United States, the Middle East, and Asia.

Polyethylene (PE)

Polyethylene is the most widely produced plastic, and it is used in packaging, construction, automotive, and other applications. Global polyethylene production capacity is estimated to be over 120 MMTPA [15]. Leading polyethylene producers include the United States, China, the Middle East, and Europe.

Polypropylene (PP)

Polypropylene is a versatile plastic used in packaging, automotive parts, textiles, and consumer goods. Global PP production capacity is estimated to be around 100 MMTPA [16]. Major PP producers include the United States, China, the Middle East, and Europe.

Aromatics (BTX)

Aromatics are important petrochemicals used in producing plastics, synthetic fibers, rubber, and pharmaceuticals. Global benzene production capacity is estimated to be around 50–60 MMTPA. Key benzene-producing regions include the United States, the Middle East, Asia, and Europe. Global toluene and xylene production capacities are estimated to be around 40–50 MMTPA [17]. Major BTX producers include the United States, China, the Middle East, and Asia.

Methanol

Methanol is a feedstock for producing chemicals, plastics, and fuels. Global methanol production capacity is estimated to exceed 100 MMTPA [18]. Leading methanol producers include China, the United States, the Middle East, and Europe.

Ammonia

Ammonia and urea are essential fertilizers produced from natural gas and nitrogen. Global ammonia production capacity is estimated to be over 200 MMTPA [19].

Significant ammonia capacity is concentrated in regions with ample fertilizer and petrochemical industries, such as China, the United States, Russia, and the Middle East.

The figures mentioned above are approximate. New projects and expansions can impact global petrochemical capacities over time. Market conditions also affect it. Therefore, it is essential to note that the petrochemical industry is dynamic, with capacities constantly changing due to factors such as demand growth, technological advancements, and investment decisions by industry players.

3.2.2 Region-Wise Major Petrochemical Complexes/Plants

Petrochemical capacities vary across regions and countries. The major players are in the United States, China, the Middle East, and Europe. Region-wise the products manufactured and capacities are varying. Capacities vary by region due to feedstock availability, investment climate, infrastructure, and demand dynamics. The following paragraphs discuss region-wise major players and products manufactured.

Middle East Region

The Middle East region is one of the major hubs for petrochemical production. The countries which have invested heavily in petrochemical complexes are Saudi Arabia, Qatar, and the United Arab Emirates. These countries have abundant hydrocarbon resources and the cost of feedstocks is low. Jubail and Yanbu in Saudi Arabia, Ras Laffan Industrial City in Qatar, and Ruwais in the UAE are the major petrochemical complexes in this region. Middle East mainly produces ethylene, propylene, PE, PP, methanol, and ammonia. Saudi Basic Industries Corporation (SABIC), Saudi Aramco, Qatar Petroleum, and Abu Dhabi National Oil Company (ADNOC) are some of the major producers.

North American Region

In the North American region, the United States is a major player in the petrochemical industry. In the USA the Petrochemical complexes are located in Texas, Louisiana, and Alberta. These complexes contribute significantly in the global capacities. These areas also have abundant shale gas resources, which have increased ethylene and derivative capacities.

Multinational companies like Dow, ExxonMobil, Chevron Phillips Chemical, Lyondell Basell, and Westlake Chemical are operating in this region. The products manufactured by these companies are ethylene, propylene, PE, PP, styrene, and other basic petrochemicals.

Asia-Pacific Region

In this region the growth of petrochemicals production is driven by countries like China, India, South Korea, Taiwan, Singapore, and Japan. These markets are witnessing rapid industrialization, urbanization, and rising consumer demand. China is one of the largest consumer and producer of petrochemicals in this region. The capacities in China are also expanding continuously. Some of the Chinese petrochemical complexes are largest in the work. These petrochemical complexes are located in Shanghai, Tianjin, and Ningbo.

China has a major capacity to produce ethylene, PE, PP, and other chemicals. This region has some of the prominent Petrochemical hubs like Zhangjiagang (China), Ulsan (South Korea), Yokkaichi (Japan), and Jurong Island (Singapore).

This region also has the capacity to produce many other petrochemical products like olefins, aromatics, polymers, and specialty chemicals.

European Region

Germany, the Netherlands, Belgium, France, and the UK are some of the major countries manufacturing petrochemicals in Europe. The petrochemical complexes in Europe are integrated with refineries. They have a benefit because of proximity to major consumer markets. It is also considered as one of the mature markets.

The major products offered by this region are ethylene, propylene, PE, PP, aromatics, and specialty chemicals. European petrochemicals suppliers have a focus on high value-added products, polymers, and specialty chemicals.

Some of the major players in this region are BASF, INEOS, Dow, Lyondell Basell, and Total Energies. European countries also have concerns for sustainability, therefore their efforts are towards circular economy initiatives in petrochemical production.

Latin American Region

Countries like Brazil, Mexico, and Argentina have petrochemical complexes. Some of the major petrochemicals facilities are in Triunfo and Camaçari in Brazil, and Altamira and Tampico in Mexico. Some of the major companies operating in this region are Braskem, Petrobras, Mexichem, and Alpek.

Some of the basic petrochemicals like ethylene, propylene, PE, PP and specialty chemicals are produced in this region. The facilitating factors for growth of this industry in the region are access to feedstock, market demand, and investment climate.

African Region

This region is witnessing the investment in infrastructure and growing demand for petrochemical products. Petrochemical industry is getting developed in countries like Egypt, Nigeria, Algeria, etc. One can find petrochemical hubs in places like Sasolburg (South Africa), Alexandria (Egypt), and Port Harcourt (Nigeria).

These petrochemical complexes produce basic chemicals like ethylene, propylene, PE, and PP. They are also into fertilizers and other downstream products. The growth of this industry in the region is because of the availability of natural resources, and infrastructure development. The government of the respective countries also are focusing their efforts on economic diversification.

3.2.3 Petrochemical Plant Capacity Utilization

This term represents the extent to which a plant's production capacity is utilized over a given period. It is calculated as the ratio of actual production output to the maximum possible output the plant can achieve under ideal operating conditions, multiplied by 100 to express it as a percentage.

Some of the key aspects of capacity utilization are discussed below.

1. **Optimization and Monitoring:** This helps in identifying inefficiencies, addressing bottlenecks, and optimizing operations in petrochemical plants. Capacity utilization can be improved through investment in technology, enhancing process efficiency, optimizing production scheduling, and aligning production with market demand.
2. **Performance Comparison and Benchmarking:** The goal of any petrochemical plant is to achieve high-capacity utilization to maximize efficiency and profitability. It is observed that the average utilization rates for well-established and efficiently operated petrochemical plants range from 80% to 95% of their production capacity.

   Performance comparison helps the petrochemical companies to evaluate their operational efficiency and help in identifying areas of improvements for optimization. The capacity utilization is compared with industry average/ peer group performance rates. It is helpful in assessing competitiveness and identifying opportunities for improvement.
3. **Factors Impacting Capacity Utilization:** The factors affecting capacity utilization of petrochemical plants are market demand, efficiency in production, maintenance schedules, and state of economy. Regulatory compliances, supply chain disruption, seasonal variations, and industry-specific factors also affect capacity utilization.
   a. **Market Demand:** Market demand for petrochemical products affects production levels and capacity utilization. Decrease in demand may result in product cuts and lower capacity utilization, while increase in demand may lead to increased production and higher capacity utilization rates. The producers adjust production levels in response to reduced demand or oversupply conditions.
   b. **Efficiency in Production:** This refers to efficiency of plant operations, equipment reliability, process optimization, and workforce productivity. Efficient plant operations contribute to achieving higher capacity utilization rates. Higher capacity utilization can be attained through investments in technology upgrades, process improvements, and workforce training.
   c. **Plant Downtime and Maintenance:** The production capacity may get reduced because of planned maintenance shutdowns, and equipment repairs. Sometimes unplanned downtime due to technical issues or maintenance backlog can also reduce production capacity and lower utilization rates. To minimize production disruptions and optimize asset performance it is suggested to have proactive maintenance and a planned shutdown schedule.

d. **Conditions of Economy**: The production and capacity utilization may be affected by economic factors like raw material prices, energy costs, exchange rates, and market competition. Recession in the economy may lead to lower utilization rates because of reduced demand and profitability concerns.
e. **Compliance to Regulations:** This refers to environmental regulations, safety standards, and product quality requirements. Regulatory compliance may impose constraints on production operations and affect utilization rates.
f. **Disruptions in Supply Chain:** The shortages of raw materials, bottlenecks in transportation, or any other logistics challenges, can limit production and affect capacity utilization.
g. **Seasonal Fluctuations in Demand:** The seasonal and cyclical fluctuations in demand of end-user industries may result in fluctuations in production levels and capacity utilization throughout the year.
h. **Factors Related to Industry**:
Capacity Utilization vary among different segments of the petrochemical industry (e.g., ethylene, polyethylene, aromatics). It also varies for an individual facility based on product demand, market competitiveness, and technology advancements. The plant age, technology, product mix, and market dynamics also affect the utilization rates.

Thus, the capacity utilization of petrochemical plants is affected by a combination of market dynamics, operational factors, economic conditions, and regulatory requirements. Appropriate production planning and efficient operations management will help achieve optimum capacity utilization. Utilization also may fluctuate over time and across the industry. Continuous monitoring, analysis, and optimization efforts are necessary for petrochemical companies to achieve high-capacity utilization levels and maximize their operational performance and competitiveness.

3.3 TECHNOLOGY IN PETROCHEMICAL PLANTS

Operations and efficiency of petrochemical plants depend upon the plant technology. Plant technology optimizes the cost and processes. Technology also plays an important role in safety and environmental compliance. Some of the key interfaces of technology in petrochemical plants are discussed in the following paragraphs.

Process Automation and Control

The real-time process monitoring, process control, and process optimization are ensured by Distributed Control Systems (DCS) and Supervisory Control and Data Acquisition (SCADA). Mathematical models and predictive analytics algorithms used in APC in the above systems optimize process parameters, increase efficiency, and minimize energy consumption. To automate equipment operation, valve control, and batch processing integrated automation systems are used. It also improves reliability of the processes and reduces human error.

Sensors and Instrumentation

To measure some of the key process variables such as temperature, pressure, flow rate, level, and composition advance instrumentation and sensors are used. The data is collected through wireless sensor networks and Industrial Internet of Things (IIoT) technologies. It enhances the data collection, analysis, and decision-making capabilities. The product quality, impurities, and composition are monitored using online analyzers on a real-time basis. The real-time monitoring enables adjust the required parameters to maintain product specifications.

Optimization and Modeling of Processes

The petrochemical processes are modeled using process simulation software. It helps in optimizing reactor design, catalyst selection, and operating conditions. Fluid flow, heat transfer, and chemical reactions are simulated through computational fluid dynamics (CFD), which improves reactor performance and energy efficiency. To optimize plant productivity and profitability production schedules, inventory management, and resource allocation are optimized through advance optimization soft wares.

Risk Management of Safety Systems

The petrochemical plants need fire and gas detection systems, video surveillance, perimeter security, plant safety, and emergency response capabilities. Safety instrumented systems (SIS) and emergency shutdown systems (ESD) ensure safe operations. These systems automatically initiate shutdown procedures in response to abnormal conditions or safety hazards. To identify and mitigate safety risks and process hazards risk assessment tools like hazard and operability studies (HAZOP) and failure modes and effects analysis (FMEA) are used.

Sustainability and Energy Efficiency

Energy consumption may be monitored using energy management system (EnMS). It will help to identify inefficiencies and implement energy-saving measures like heat integration, cogeneration, and recovery. Energy-intensive operations may be optimized through process optimization software. It will help reduce greenhouse gas emissions, and improve environmental performance. To capture CO_2 emissions carbon capture and utilization (CCU) technologies are used. The captured CO_2 in petrochemical processes is used for reuse or storage which contributes to mitigating greenhouse gas impact.

Connectivity and Digitalization

Real-Time plant data can be monitored from remote places a using mobile applications. Alerts received through such applications help operators to make informed decisions remotely. The appropriate information and communication technologies (ICTs) may be used for predictive maintenance, performance optimization, and scenario analysis. The data storage, analytics, and collaboration across geographically distributed sites are managed through cloud computing facilities. It enhances decision-making and knowledge sharing.

To drive innovation, improve operational efficiency, enhance safety, and achieve sustainable growth technology integration is essential in the petrochemical industry. Industry challenges and opportunities are met through focused investment in research and development, technology collaborations partners, and adoption of best practices. These efforts also help in ensuring competitiveness in the industry.

3.3.1 Manufacturing Processes (Petrochemicals and End Products)

The manufacturing processes of petrochemicals have various stages; it involves converting raw materials to intermediate products and further converting intermediate products to final products. The conversion processes involve various chemical reactions and engineering processes which convert raw materials like crude oil, natural gas, and NGLs (natural gas liquids) into intermediate and final petrochemical products. Figure 3.3 shows the diagram of typical manufacturing processes from raw material to end products. A typical manufacturing process involved in petrochemical production is discussed below.

Procurement and Preparation of Feedstock

The main feedstocks which are used as a raw material to make petrochemicals are primarily crude oil and natural gas. Manufacturing process starts with procuring and preparing feedstocks, which could be crude oil, NGLs, naphtha, ethane, propane, and butane. Intermediate feedstocks are supplied by refineries, natural gas processing plants, or may be imported sources.

Refining of crude oil in oil refineries produces naphtha, gas oil, and other hydrocarbon fractions which are used as intermediate feedstocks. While, Natural gas processing separates natural gas liquids (NGLs) such as ethane, propane, and butane from methane.

Primary Petrochemical Production (Cracking)

The primary petrochemical production stage involves converting feedstocks into basic petrochemicals through various chemical processes. Steam cracking primarily breaks down hydrocarbons into simpler molecules, such as ethylene, propylene, and butadiene, and aromatics like BTX. Steam cracking is the most common method, involving heating hydrocarbon feedstocks in the presence of steam at high temperatures (800–900°C) to induce thermal decomposition.

FIGURE 3.3 Manufacturing Process for Petrochemical Products.

Thus, cracking processes primarily yield olefins such as ethylene, propylene, and butadiene, key building blocks for various petrochemical products. Separation techniques such as distillation, absorption, and fractional crystallization are used to separate and purify the olefins from other by-products.

Other processes include catalytic cracking, reforming, dehydrogenation, and alkylation to produce additional petrochemicals and intermediates. It uses catalysts to facilitate the cracking reactions.

Intermediate Chemical Production

Further processing of basic petrochemicals produces intermediate chemicals through chemical reactions. Examples include ethylene oxide, ethylene glycol, styrene, methanol, and acrylonitrile. Various chemical synthesis processes such as oxidation, hydrogenation, halogenation, and polymerization produce intermediate chemicals with specific properties.

These Intermediate chemicals are produced through additional chemical reactions involving olefins, aromatics, or other feedstocks. It serves as building blocks for synthesizing specialty chemicals, plastics, resins, and other downstream products.

Aromatics Production

Aromatics such as BTX are produced from feedstocks rich in aromatic compounds or through processes like catalytic reforming. Aromatics are precursors for making plastics, synthetic fibers, solvents, and other chemicals.

Manufacturing of Polymers

Polymers are produced through polymerization processes. The polymerization process transforms intermediate chemicals and basic petrochemicals into polymers. Polymers are long-chain molecules with repeat units. Olefins and other monomers are transformed into polymers. Standard polymerization techniques are briefly discussed below.

1. **Polyethylene (PE):** Polymerization of ethylene with high pressure or low pressure.
2. **Polypropylene (PP):** Catalysts are used during polymerization of propylene.
3. **Polyvinyl chloride (PVC):** PVC is produced by polymerization of vinyl chloride monomers.
4. **Polystyrene (PS):** It is produced through polymerization of styrene monomers.
   The processing of polymers can be done into various forms, such as pellets, resins, fibers, films, and sheets.

Polyethylene (PE) and polypropylene (PP) are produced through polymerization of ethylene and propylene. Catalysts are used in this polymerization process. Reaction conditions are tailored to control molecular structure and properties. Additional polymers such as PVC, PS, and polyethylene terephthalate (PET) are produced through various polymerization methods.

Refining and Purification of Products

Impurities and contaminants in the petrochemical products are removed through refining and purification processes. This process also separates by-products. Some of the refining and purification techniques are distillation, solvent extraction, crystallization, and filtration. These processes are used to achieve desired product specifications and purity levels.

Formulation and Packaging of Petrochemical Products

Formulation converts petrochemical products into specific grades as per customer requirements. Packaging and labeling processes ensure proper petrochemical product storage, handling, and transportation to end-users and downstream industries.

Downstream Manufacturing

Downstream manufacturing refers to finished products or end-use products manufacturing using polymers or intermediate petrochemicals as raw material. Finished products examples are plastic articles, packaging materials, fibers, films, automotive components, construction materials, and consumer goods. Manufacturing processes used for finished products are injection molding, blow molding, extrusion, casting, calendaring, and spinning. These manufacturing processes use advanced technologies, process controls, and engineering solutions throughout the processes to optimize efficiency, safety, and product quality. Focused and continuous efforts in Research and Development (R&D) lead to innovation in petrochemical manufacturing. R&D also improves process economics, environmental sustainability, and product performance. Safe and sustainable petrochemical production requires complying with regulatory standards and adopting the best practices of the industry.

Testing and Quality Control

Total Quality Management system is adopted throughout the manufacturing process to guarantee product consistency, purity, and compliance with specifications. Advance Testing and analysis techniques such as chromatography, spectroscopy, rheology, and mechanical testing are used to assess product quality and performance characteristics.

Environmental Considerations and Waste Management

The manufacturing of petrochemicals generate various by-products, wastes, and emissions. It requires proper management to minimize environmental impacts. To mitigate environmental pollution recycling, reuse, energy recovery, and waste treatment are employed.

The above-mentioned manufacturing processes show the steps involved in the production of petrochemicals, from raw material procurement to the manufacturing of finished products. The processes can be optimized through focused and continuous improvement efforts, and technological advancements. Optimized processes enhance the efficiency and sustainability of petrochemical manufacturing operations.

3.3.2 Petrochemicals: Quality Issues

Petrochemical products quality refers to products' safety, efficiency, and sustainability. The quality of petrochemical products is dependent on many factors like raw material quality and selection of manufacturing processes. Final product testing must ensure that the products meet the necessary standards and specifications for their intended applications. As such the petrochemical industry is highly regulated, which requires strict compliance to quality standards. It is also essential to meet environmental and safety requirements, maintain equipment integrity, and produce high-grade end products. To preserve and enhance the quality focused efforts for continuous improvement and strict compliance to international standards and regulations are necessary. The key quality aspects are:

1. **Raw Material Quality/ Quality of Feedstock**
    a. **Chemical Composition and Purity:** The quality of petrochemicals is affected by the quality of crude oil or natural gas. Inconsistency in quality of crude oil and natural gas impacts the yield and efficiency. Impurities can affect the performance and characteristics of petrochemical products. Petrochemical products should have the correct chemical composition with least impurities. Consistent supply of raw materials from reliable sources ensures uniformity in product quality.
    b. **Contaminants:** The quantum of contaminants like sulfur, metals, nitrogen, and other unwanted chemicals cause corrosion, fouling, and catalyst poisoning in downstream processes.
2. **Management of Catalyst**
    a. **Quality of Catalyst:** Efficiency of chemical reactions is very much dependent on quality of catalysts. It also affects the product quality and yield.
    b. **Replacement and Regeneration:** Replacement and Regeneration on a regular basis are required to maintain activity and selectivity.
3. **Physical Properties of Products and Consistency of Products**
    a. **Density:** Accurate measurement and control of density.
    b. **Viscosity:** Ensuring the correct flow characteristics.
    c. **Boiling and Melting Points:** Precise determination of these points for proper application.
    d. **Refractive Index:** Important for specific applications, particularly in polymers.
    e. **Batch-to-Batch Consistency:** Maintaining uniform quality across different production batches.
    f. **Stability:** Ensuring chemical and physical stability over time.
4. **Process Control and Monitoring**
    a. **Advanced Process Control:** To optimize the production processes, advanced process control systems are used. It ensures in maintaining the process variables within desired range and improves process efficiency as well as product quality. Product quality and consistency can be achieved through proper control of temperature, pressure, and reaction time.
    b. **Real-Time Monitoring:** Continuous monitoring of key parameters (temperature, pressure, and flow rates) ensures the process remains within specifications, reducing the risk of defects.
    c. **Hazard Analysis**: Identifying potential hazards in production and implementing measures to mitigate risks.

5. **Product Quality Control and Analytical Testing Methods**
    a. **Compliance for Specifications:** This means that the finished products meet stringent specifications with respect to purity, composition, and other properties (e.g., specifications related to octane number for gasoline, specifications like polymer grades for plastics).
    b. **Analysis and Testing:** To meet the quality standards comprehensive testing is done (e.g., chromatography, spectroscopy). It helps to detect and quantify impurities and contaminants which may affect product performance and safety.
    c. **Spectroscopy:** Using techniques like NMR, IR, and UV-Vis for quality control.
    d. **Chromatography:** Employing methods like GC and HPLC to analyze chemical composition.
    e. **Mass Spectrometry:** This is required for precise molecular analysis.
6. **Performance**
    a. **Performance for Functionality:** This is required to check how well the petrochemical product performs for its intended use. It may be related to fuel efficiency in the case of gasoline or lubrication properties for oils.
    b. **End-Use Specifications:** The product should meet the specific requirements for various end-use applications like tensile strength in polymers.
7. **Documentation and Traceability**
    a. **Records of Each Batch**: It is critical to maintain documentation and records of every batch. It is required to ensure traceability and accountability.
    b. **Quality Control Records**: The regulations require maintaining quality records. It also helps in quality assurance.
8. **Managing the Supply Chain**
    It is with respect to timely supply of raw materials, equipments and services, logistics, and storage. Some of the petrochemical products are affected because of temperature and humidity, proper storage is needed to ensure that the product quality does not get affected. Suitable packaging is required to prevent contamination and degradation.
9. **Customer Satisfaction**
    a. **Customization**: Ability to tailor products to specific customer needs and specifications.
    b. **Technical Support:** Providing customers with necessary technical information and support.
    c. **Customer Feedback:** To ensure customer satisfaction a proper feedback system about product quality should be implemented.
10. **Sustainability, Innovation, and Technology**
    a. **Green Chemistry**: Adopt environmentally friendly chemical processes and develop sustainable products.

b. **R&D**: Investment in research and development to innovate and improve product quality, improve processes, and develop new products.
c. **Digitalization:** Utilizing digital technologies (e.g., IoT, AI, machine learning) for predictive maintenance, process optimization, and quality monitoring.

11. **Safety and Environmental Standards/Regulatory Compliance**

Compliance for Safety: The products should be free of any harmful and toxic substances. If the material is flammable, it should be classified accordingly for proper handling. Safety compliances with respect to proper handling, storage, and transport should be ensured.

Standards for Environment: The norms related to emission controls (for CO_2 and NO_x, and SO_x) during production of petrochemicals are to be complied with so that it does not harm the environment and public health. Also, it is required to handle the hazardous material properly in terms of its handling, storage, and disposal.

12. **Quality Management Systems**

a. **ISO Standards**: Internal and external audits are to be conducted to ensure quality standards. Implementation of ISO 9001 and other relevant standards are required to ensure consistent quality management practices.
b. **Continuous Improvement**: Adoption of Six Sigma and Lean methodologies to continuously improve processes and product quality.
c. **Preventive Maintenance:** Timely and regular maintenance of equipment (like heat exchangers and reactors) used in various processes prevents unexpected failures and maintains process integrity.

The above-mentioned aspects help petrochemical companies to produce high-quality products which meet regulatory requirements and customer expectations. It also ensures operational efficiency and sustainability.

BOX 3.1 THE CASE OF TEFLON

One of the popular petrochemical products is polytetrafluoroethylene (PTFE), a synthetic fluoropolymer of tetrafluoroethylene. Being hydrophobic, non-wetting, high density, and resistant to high temperatures, PTFE is an incredibly versatile material with various applications. PTEF is best known for its non-stick properties. It is also popularly known as Teflon. Teflon™ is a synthetic polymer containing carbon and fluorine, making it PTFE. Teflon™ is a trademarked name for PTFE, owned by Chemours. It is neither metal nor rubber. It is also not precisely plastic. More specifically, PTFE is a thermoplastic polymer primarily characterized by a slippery surface, high melting point, and resistance to "attack" by almost all chemicals.

Key characteristics of PTFE

While a plastic molecule has a carbon chain with hydrogen atoms attached, the hydrogen atoms in a PTFE structure are replaced with fluorine atoms. It drastically alters the material's properties, and PTFE, like other fluoroplastics, has unique properties:

Benefits of PTFE

PTFE also has a wide operating temperature range of –30°C to +200°C, a hardness range of 70 to 85 IRHD, and is simple to clean. Given these positive characteristics, PTFE is a relatively inexpensive elastomeric option for seals and "O" rings. PTFE is available in four forms.

1. **Granules**: These are used to make various components like rods, sheets, rings, etc.
2. **Fine Powder:** Used for pipes, thread sealant tapes, tubing, and valve coatings in chemicals, petrochemicals, sugar factories, etc.
3. **Dispersions:** It is a mixture of sixty percent Teflon and forty percent water, used for coated laminates, glass fiber, etc.
4. **Industry Finishes:** Used for coating vessels used for various chemical processes.
5. **Micropower:** Used for paints, inks, lubricants, engine oils, etc.

All the above forms have many combinations.

The quality of this product varies from supplier to supplier. The major player supplying this product is DuPont, which is now Chemourse. Quality is critical for buyers (i.e., user industry), such as components and parts used for aircraft manufacturing, satellite/space applications, defense applications, etc. Such buyers have the practice of having specifications written during the contract (Spec-In) with the suppliers of components/parts wherein raw material, i.e., Teflon to be used, should be from Chemourse (product Specs). The component or part manufacturers (if the end products to be made from this raw material are not very critical from the perspective of safety issues) sometimes play with this contract condition by buying small quantities and getting COA (certificate of analysis) and QC (quality control) Test certificate from Chemourse and rest from other suppliers as the price of other suppliers products is one-third of Chemourse. This example shows how the quality of products from different suppliers varies.

3.4 PREPARING FEASIBILITY REPORTS OF PETROCHEMICAL COMPLEXES/PLANTS: AN ILLUSTRATION

Feasibility reports for petrochemical plants are comprehensive assessments to evaluate the technical, economic, financial, environmental, and regulatory aspects of establishing or expanding a petrochemical facility. It is typically prepared by multidisciplinary teams comprising engineers, economists, financial analysts, environmental specialists, and regulatory experts. The findings and recommendations of the feasibility study inform stakeholders, investors, lenders, and project sponsors about the feasibility, risks, and opportunities associated with the proposed petrochemical project, facilitating informed decision-making and project development. The feasibility reports play an important role for decision makers to evaluate the techno-economic viability of the project. It also helps identify risks associated with the proposed project.

The outline of the key components typically included in feasibility reports for petrochemical plants are as follows.

1. Executive Summary

It typically contains a brief overview of the project, which includes project objectives, scope of project, key findings, and recommendations.

2. Introduction

This section discusses the rationale for the project, drivers of market demand, supply chain dynamics, and strategic importance.

3. Market Feasibility/Analysis

Market feasibility study or market analysis requires demand-supply of various products, market trends, competition, market share of various players, etc. It also contains potential challenges and opportunities in the target market. Demand-supply analysis suggests opportunities for the new player in the industry to set up the new capacities of petrochemical plant along with the products to be made in the new plant. The petrochemical industry marketing experts, industry association, market research firms, etc. help in undertaking market feasibility analysis.

Market feasibility studies for petrochemical plants typically contain the following sections.

3.a Analysis of Market Demand

In this section, current and future demand for various petrochemical products is estimated. To forecast the demand for each segment of products, i.e. propylene, PE, PP, aromatics, ethylene, etc. Demand is further forecast for each user industry like automotive, consumer goods, cosmetics, construction, packaging, etc. The variables used to forecast the demand are population growth, economic growth, consumption trends, industrial growth, level of urbanization, etc.

3.b Market Segments, Entry Strategy, and Competitive Analysis

To segment the market one can use parameters like product types, applications, end-users, geographic regions, etc. Based on the capabilities of the plant and product portfolio target market segments with growth potential, profitability, and strategic fit are identified. Market entry strategy is planned keeping in view the factors like market attractiveness, competitive intensity, regulatory compliance, and risk mitigation. While planning the market entry strategy it is also required to take into account targeted market segment, pricing policy, promotion campaigns, and distribution channels. Market entry options are evaluated based on strategic objectives and resource capabilities. Further the entry strategy will also get influenced by the factors like green field projects, joint ventures, acquisitions, partnerships, or alliances. Regulatory risks, compliance costs, and market access barriers may also affect market entry.

The competitive analysis of the petrochemical industry in the target market involves identifying existing players, their market shares, strengths, weaknesses, strategies, and barriers to entry. Thus, the multinational corporations, national companies, and regional players competitive position is analyzed. The analysis of competitive position will help in identifying competitive advantages and differentiation opportunities for the proposed petrochemical plant. Technology innovation, product differentiation, cost leadership, niche markets, etc. are some of the competitive advantages for the new players.

3.c Risk Assessment and Opportunities

Risk assessment and opportunities are with respect to uncertainties associated with market conditions. It may include demand volatility, price fluctuations, currency exchange rates, political instability, and macroeconomic factors. Risk mitigation strategies and contingency plans are developed to address market risks.

The unmet needs, emerging trends, technological innovations, or changing consumer preferences are identified for potential opportunities. Further it is suggested to evaluate strategic growth options like diversification, product innovation, value chain integration, or market diversification into new regions or applications.

3.d Analysis of Supply Chain

Supply Chain analysis involves analysis of availability, reliability, and cost competitiveness of feedstock (i.e. crude oil, natural gas, NGLs, and other hydrocarbons) sources. It is also suggested to analyze supply chain dynamics sources of supply, logistics, distribution channels, and access to the market. The risk involved in managing the supply chain may be because of geopolitical factors, infrastructure constraints, and market disruptions. The relationship with the supplier's and procurement strategies are evaluated to mitigate the supply chain risk in order to ensure reliable access to raw materials.

3.e Regulatory Environment and Policy Environment

This involves analysis of policies, standards, regulatory frameworks, and incentives affecting the petrochemical industry. The promoters are required to understand the licensing procedures and various permissions required to set up and operate the petrochemical plants. They are related to environmental regulations, trade policies, taxation, subsidies, and investment incentives.

3.f Engagement of Stakeholders

Since the new petrochemicals plant is being planned it is advisable to engage with key stakeholders to get their insights. The stakeholders with whom one must engage are customers, suppliers, government agencies, industry associations, and community groups. The relationship building with them will address concerns related to market feasibility.

4. Technical Feasibility/Analysis

The technical feasibility of a petrochemical plant refers to designing, constructing, and operating the plant to produce desired petrochemical products efficiently and reliably. It helps evaluate technical aspects of process technology, engineering design, equipment selection, feedstock availability, site suitability, and infrastructure requirements. This analysis requires expertise from process engineers, chemical engineers, mechanical engineers, electrical engineers, and other technical specialists. It is also suggested to collaborate with technology licensors, equipment suppliers, and industry experts for technical analysis.

Some of the key modules to evaluate the technical feasibility of petrochemicals project are discussed below.

4.a Selection of Technology for Process

This refers to selection and evaluation of appropriate process technologies to produce desired petrochemical products, such as ethylene, propylene, PE, PP, aromatics, and specialty chemicals. The factors to be looked at for selecting the process technologies are process efficiency, product quality, scalability, environmental impact, energy consumption, and capital costs. The suitable process routes like steam cracking, catalytic cracking, reforming, polymerization, and other chemical synthesis methods are also assessed.

4.b Layout and Plant Configuration

The layout and plant configuration design is done based on process requirements, equipment specifications, and site constraints. Further the factors like material handling, utilities integration, safety considerations, and future expansion are kept in mine while planning layout and plant configurations. The process flow, arrangement of equipments, and interconnections are shown using process flow diagrams (PFDs) and piping and instrumentation diagrams (P&IDs).

4.c Scale and Capacity of Production

The selection of process technology and equipment shall be based on its scalability to take care of future capacity modifications and expansion. The decision related to production capacity of the petrochemical plant is based on market demand, planned investment, and economies of scale.

4.d Selection and Design of Equipment

The appropriate machinery and equipment's are to be selected for various unit operations. The machinery and equipments used are distillation columns, reactors, heat exchangers, compressors, pumps, and control systems. While selecting the equipment its reliability, performance specifications, compatibility with process conditions, and maintenance requirements are looked at. Further the selection should also use the parameters like codes, standards, regulations related to equipment design and fabrication.

4.e Supply and Logistics of Feedstock

Availability and logistics of feedstock (i.e. crude oil, natural gas, NGLs, and other hydrocarbons) is one of the critical requirements for petrochemicals projects. One needs to evaluate feedstocks' availability, quality, and cost. The project requires assessing the sourcing and logistics options for feedstock, its storage facility, and resilience to the supply chain. Further the feedstock availability is analyzed with respect to its variability, market dynamics, seasonality, and geopolitical risks.

4.f Infrastructure and Utilities

The infrastructure for the project requires site preparation, connecting roads, connections to utilities, wastewater treatment, and environmental controls. The requirements for utilities are for steam, electricity, water, and cooling systems to support plant operations. The infrastructure should also be looked at from the perspective of safety requirements, response for emergency, and regulatory compliances.

4.g Risk Management for Process Safety

The risks associated with petrochemical processes are fire, explosion, chemical release, and toxic exposure. These risks are identified and mitigation strategies are planned for potential hazards. The strategies to mitigate risks are (i) implementation of process safety management systems, hazard analysis techniques (e.g., HAZOP, FMEA), and safety instrumented systems. These Systems help in preventing accidents and protect personnel, assets, and the environment.

4.h Environmental Regulations

It is required to comply with environmental regulations. It is with respect to managing air emissions, proper water discharges, management of waste generation, and optimized resource consumption. One needs to develop environmental management plans and plan pollution prevention strategies to comply with regulatory requirements related to the environment.

5. Financial Analysis

Financial analysis requires projecting the revenue potential based on market potential, profitability, and return on investment (ROI) of the proposed petrochemical project. In the financial feasibility analysis revenue estimates are projected based on sales volumes and price assumptions. To estimate the profit margins, cost structures i.e. capital cost and operating cost are worked out. Various financing options like debt, equity, subsidies, incentives, etc. are looked at for capital cost analysis. Financial models are developed to forecast cash flows, revenues, expenses, and profitability over the lifecycle of a project. It also helps in sensitivity analysis. It is advisable to evaluate financial risk using factors like market volatility, currency exchange rates, interest rates, and inflation.

Further, the operating expenses for the project are worked out. It includes raw material costs, running cost of utilities, labor cost, maintenance cost, and overhead cost. In the financial feasibility analysis project economics is worked out which requires calculating financial parameters such as net present value (NPV), internal rate of return (IRR), payback period, and profitability index.

6. Economic Analysis

Economic analysis is undertaken to find out cost and benefit of the petrochemical project. The cost analysis includes requirements for capital investment for land acquisition, construction of plant and buildings, procurement, and installation of equipment, and startup costs. The benefit analysis includes analysis of economic benefits in the context of developing the region's economy, generating the employment in the region, opportunities for ancillary industry, etc. One can also look at the overall benefit of the project to the society and government in this analysis.

7. Environmental and Regulatory Analysis

All the requirements related to environmental compliances and regulatory compliances are identified for the project implementation and operations. Environmental impact analysis (EIA) is carried out to identify environmental risks and the ways to mitigate those risks. Further HSE issues are to be looked and appropriate systems/plans are to be developed, HSE may also include emergency response plans, emission control, and waste management.

8. Project Implementation

A plan to implement the project is developed giving details about important milestones, timelines, and deliverables and resource requirements. Resource requirements related to equipments, machinery, manpower is worked out to execute the project without any constraints. Project management tools and techniques like networking techniques i.e. CPM and PERT are used to monitor the progress of the project work. Project risk mitigation strategies should also be in place to successfully implement the project.

3.4.1 Illustrative Case Study of a Petrochemical Project[1]

Proposed Project: Ethylene Production Plant

1. **Location:** Gulf Coast, United States.
2. **Capacity:** 1,000,000 metric tons per year (MMTPA).
3. **Feedstock:** Ethane sourced from natural gas processing.
4. **Technology:** Steam cracking process.
5. **Project Timeline**: Construction period of 3 years, followed by operation over 20 years.
6. **Financial Assumptions:**
   - A Capital Costs: Total Capital Investment: $1.5 billion
     Breakdown of Capital Cost:
     - i Engineering and Design: $100 million
     - ii Equipment Procurement: $800 million
     - iii Construction: $500 million
     - iv Contingency: 10% of total capital investment
   - B Operating Costs:
     - i Raw Material Costs: $300 per metric ton of ethylene produced
     - ii Utilities (steam, electricity): $50 per metric ton of ethylene produced
     - iii Labor and Maintenance: $30 million annually
   - C Revenue:
     Average Selling Price of Ethylene: $800 per metric ton
     Sales Volume: 1,000,000 metric tons per year
   - D Financing:
     - i Debt Financing: 70% of total capital investment at an interest rate of 5%
     - ii Equity Financing: 30% of total capital investment
   - E Financial Analysis:
     - i Capital Investment:
       - a. Total Capital Investment = $1.5 billion
       - b. Debt Financing: $1.05 billion
       - c. Equity Financing: $450 million
     - ii Operating Expenses:
       - a. Raw Material Costs: $300 million per year
       - b. Utilities: $50 million per year
       - c. Labor and Maintenance: $30 million per year
   - F Revenue:
     Annual Revenue = Selling Price * Sales Volume = $800 * 1,000,000 = $800 million

- G Profitability Analysis:
  - i Operating Profit = Revenue – Operating Expenses = \$800 million – (\$300 million + \$50 million + \$30 million) = \$420 million
  - ii EBITDA (Earnings Before Interest, Taxes, Depreciation, and Amortization) = \$420 million
  - iii Net Income = EBITDA – Depreciation – Interest Expenses – Taxes
- H Cash Flow Analysis:
  - i Calculate annual cash flows from operations (revenue minus operating expenses), debt service (principal and interest payments), and equity returns.
  - ii Use discounted cash flow (DCF) analysis to determine the project's net present value (NPV) and internal rate of return (IRR).
  - iii Assess the payback period and profitability index to evaluate the investment's financial attractiveness.
- I Sensitivity Analysis:

  Conduct sensitivity analysis to assess the impact of changes in key variables such as ethylene prices, raw material costs, and operating expenses on project economics. Evaluate the project's resilience to adverse scenarios and identify sensitivity thresholds.

  Based on the financial analysis, stakeholders can make informed decisions regarding the economic viability and feasibility of the proposed ethylene production plant. In their decision-making process, they may consider factors such as projected returns, investment risks, market conditions, regulatory considerations, and strategic objectives. Continuous monitoring and reviewing financial performance throughout the project lifecycle is essential to ensure the project's success and maximize value for investors and stakeholders.

NOTE

1 Please note that all figures are assumed and are for illustrative purpose only.

REFERENCES

1. Reliance Industries Limited, www.ril.com, Retrieved on May 25, 2024 from www.ril.com
2. Jurong Iland Petrochemical Complex, www.pcs.com.sg/singapore-essential-chemicals-complex/jurong11island/, Retrieved on May 25, 2024.
3. Texas Gulf Coast Petrochemical Complex, Offshore-Technology.com www.offshore-technology, Retrieved on May 25, 2024.
4. Texas Gulf Coast Petrochemical Complex, Gulf Coast Growth Ventures, www.offshore-technology.com/marketdata/gulf-coast-growth-ventures, Retrieved on April 21, 2024.
5. JUbail Industrial City Ptrochemical Complex, www.offshore-technology.com/data-insights/jubail-united, Retrieved on May 25, 2024.

6. Jubail Indutrial City Petrochemical Complex, www.saudikayan.com/en Retrived on 25 May 2024.
7. Nanjing Petrochemical Complex, www.offshore-technology.com/data-insights/basf-ypc-nanjing19complex-china/ Retrieved on 25 May 2025.
8. Nanjing Petrochemical Complex, www.offshore-technology.com/data-insights/basf-ypc-nanjing19complex-china/ Retrieved on 25 May 2025.
9. Statista, Statista Research Department, www.statista.com/statistics/1260037/global-petrochemical-production-capacity/ Retrieved on May 28 2024.
10. Offshore Technology, Global Data's Petrochemical Database, www.offshore-technology.com/data-insights/top-ten-active-petrochemical-complexes-global/?cf-view Retrieved on May 29 2024.
11. Offshore Technology, Market Data, www.offshore-technology.com/marketdata/gulf-coast-growth-ventures-gcgv-san-patricio-complex-united-states/
12. Global Energy Infrastructure, News, Gulf-coast-growth-ventures-petrochemical project in Texas start up, https://globalenergyinfrastructure.com/news/2022/01-january/gulf-coast-growth-ventures-petrochemical-project-in-texas-starts-up/ Retrieved on May 15 2024.
13. Statista, Statista Research Department, www.statista.com/statistics/1067372/global-ethylene-production-capacity/ Retrieved on May 28 2024.
14. Statista, Statista Research Department, www.statista.com/statistics/1246689/propylene-demand-capacity-forecast-worldwide Retrieved on May 25 2024.
15. Statista, Statista Research Department, www.statista.com/statistics/1246675/polyethylene-demand-capacity-forecast-worldwide/ Retrieved on May 25 2024.
16. PR Newswire, Research & Markets, www.prnewswire.com/news-releases/global-polypropylene-industry-research-report-2023-demand-production-capacity-industry-applications-size--forecasts-2022-2028-301921768.html#:~:text=In%202022%2C%20the%20global%20polypropylene,the%20period%20of%202023%2D2028 Retrieved on May 21 2024.
17. Market Research Future, Secondary Research, Primary Research, MRFR Database and Analyst Review, www.marketresearchfuture.com/reports/aromatics-market-930 Retrieved on May 25 2024.
18. Methanol Institute, www.methanol.org/the-methanol-industry/, Retrieved on 22nd April 2024.
19. Statista, Statista Research Department, www.statista.com/statistics/1065865/ammonia-production-capacity-globally/ Retrieved on May 25 2024.

4 Marketing Aspects of Petrochemical Business

4.1 INTRODUCTION

Petrochemical products, having a derived demand have unique characteristics and applications. Given the nature of products, buyers, and end-use contexts, the milieu of petrochemical marketing largely hinges on the framework of business-to-business (B2B) marketing. This chapter starts with the discussion on the context of B2B marketing for petrochemical products (and services/solutions as applicable). Later, a brief discussion on the marketing mix framework with a special focus on digital technology-based pricing models, integrated marketing communications, and sales and distribution, followed by a unique Go-to-Market (GTM) model for petrochemical marketing have been undertaken. The chapter also includes the tenets of segmentation, targeting, and positioning (STP), solutions-based marketing approach, and digital marketing's role in petrochemicals marketing. The chapter includes relevant company-specific contexts/illustrations wherever applicable.

4.2 MARKETING OF PETROCHEMICALS: THE B2B MARKETING CONTEXT

Marketing petrochemicals in a **B2B** context shares several core principles with general B2B Marketing but also has unique characteristics due to the nature of the products and industry.

4.2.1 Core Principles of B2B Marketing in Petrochemical Business

1. **Understanding Customer Needs**
   a. **Technical Specifications**: Customers often require specific grades and compositions of petrochemicals to suit their manufacturing processes and are very particular about the same. Marketing teams must possess a deep understanding of the chemical properties, production processes, and applications of their products.

DOI: 10.1201/9781032661933-4

b. **Reliability and Consistency**: Consistent product quality and reliable supply are critical because any deviations from these have an impact on the end-product.
c. **Support Services**: Technical support, after-sales service, and customer education can be important differentiators. Educating customers about the technical advantages and applications of different petrochemical products is essential. Upcoming sections shall describe the ways petrochemical marketers' intense engagement with their customers towards this function.

2. **Complex Decision-Making Process**
Petrochemicals marketing involves typical long sales cycles that involve multiple stakeholders from various functional areas and departments.
3. **Relationship Building**
Trust and customer relationships are important features of the process.
   a. **Trust and Reliability**: Building robust relationships based on trust, reliability, and reliable product quality is central for petrochemical marketers.
   b. **Customer Retention**: Petrochemical marketers know that long-term relationships and customer retention are key to success. In addition, an emphasis on maintaining high service levels and addressing customer needs promptly is also important.
4. **Customization and Personalization**
Given the nature of business, petrochemical marketers understand the importance of customized solutions with specific technical assistance is an important feature.
   a. **Customized Solutions**: Petrochemical products often need to be customized to meet detailed technical requirements of diverse industries and clients that may have multiple applications.
   b. **Technical Support**: This involves extending personalized technical support and after-sales service to ensure optimal use of these products.

4.2.2 Unique Aspects of Petrochemicals Marketing in B2B Context

1. **Regulatory and Environmental Compliance**
The distinct nature of petrochemical products requires a set of regulations to be followed so that compliance related to sustainability, etc. can be ensured.
   a. **Compliance to Regulations**: Scrupulous adherence to environmental and safety regulations influences marketing strategies of petrochemical marketers. This adherence includes complying with international standards like Registration, Evaluation, Authorization, and Restriction of Chemicals (REACH) [1] in Europe.
   b. **Ensuring Sustainability**: This involves an accent on sustainability and eco-friendly products, which in turn also shape marketing communications and product development.

FIGURE 4.1 Petrochemical Marketing.

2. **Global Supply Chain Management**
   Given the international nature of petrochemical business involving different geographical regions, it is crucial to manage the supply chains with a global perspective.
   a. **Logistics Management**: Given the significance of timeliness in delivery and the impending impact of geopolitical factors on supplies, it is important to ensure efficient management of the global supply chain of the petrochemical industry.
   b. **Tackling Regional Differences**: Petrochemical marketers should keep in mind that their marketing strategies are tailored to regional market conditions, regulatory environments, and customer preferences.
   See Figure 4.1.

4.2.3 Petrochemicals Marketing and Some Other Important Contextual B2B Marketing Tactics

As we know, B2B Marketing involves some unique tactics that puts it apart from general marketing applicable to B2C and other products. Petrochemicals marketing also bank on such distinctive tools for best outcomes. These tools are discussed as follows:

1. **Content Marketing**
   This involves not only preparing the right content, but also levering digital platforms to leverage the power of content.
   a. **Preparing Technical Whitepapers and Case Studies**: Creating detailed technical documents that showcase product/service benefits, applications, and also writing case studies of successful applications of these. Some examples of such documents are shared later in the chapter.
   b. **Conducting Webinars and Workshops**: The petrochemical marketing team should host informative webinars and workshops to engage with potential customers and demonstrate their expertise. Such activities aid in concentrating the marketing efforts to the intended audiences.
2. **Trade Shows and Industry Conclaves**
   Such platforms are crucial meeting points of marketers and customers including potential customers.
   a. **Trade Exhibitions**: Participating in industry-specific trade shows to showcase products and network with key players are helpful for petrochemical marketers. Later in the chapter, some examples have been given of such events.
   b. **Business Conferences**: Attending and speaking at industry conferences to build brand reputation and generate leads helps the marketers. This also helps in showcasing the expertise and also networking with major petrochemical industry stakeholders including customers, consultants, regulators, and policy makers.
3. **Account-Based Marketing (ABM)**
   Specific nature of petrochemicals marketing requires that the marketers not only run targeted campaigns but also tailor-make their communications with customers.
   a. **Designing Targeted Campaigns**: This involves developing highly targeted marketing campaigns for specific key accounts that focus on their unique needs and challenges. The petrochemicals industry worldwide has a practice of assigning these tasks to dedicated Account Managers.
   b. **Crafting Personalized Communication**: Tailoring communication and marketing efforts to individual clients, with an aim to build deeper relationships, gives a competitive edge to petrochemical marketers.
4. **Digital Marketing**
   This involves leveraging technological tools to optimize search engine-based marketing efforts for the best results for petrochemical marketers. Some highlights are discussed below:
   a. **SEO (Search Engine Optimization) and SEM (Search Engine Marketing)**: These techniques aim at optimizing websites and using search engine marketing to attract relevant traffic and generate leads from relevant target markets. This also aids in streamlining digital marketing campaigns for best results. This further involves using

online platforms and social media to reach out to a wider audience and engage with these customers. One of the upcoming sections discusses this aspect in detail.
   b. **Social Media**: Leveraging professional networks like LinkedIn to connect with industry professionals and share thought leadership content.
5. **Customer Relationship Management (CRM)**
   a. **CRM Systems**: Implementing robust CRM systems to manage customer interactions, track sales activities, and analyze customer data to improve service. Petrochemical marketers also leverage the power of specialized CRM software towards this effort.
   b. **Feedback Loops**: Establishing mechanisms to gather and act on customer feedback to continuously improve product offerings and customer satisfaction. This helps make the customer engagement better.
6. **Branding and Positioning**
   a. **Quality and Reliability**: Emphasizing product quality, consistency, and reliability is the practice in the petrochemical marketing process.
   b. **Sustainability**: As all major stakeholders including final consumers are becoming aware about environmental impacts of the products they use, there is an increasing focus on environmentally friendly products and practices.

   Later in the chapter a detailed discussion on the role of branding in petrochemical marketing has been undertaken with the help of a case study.
7. **Strategic Partnerships**
   a. **Joint Ventures**: Collaborating with other companies to expand market reach or share technology. Collaborations are also helpful in co-creation of products.
   b. **Alliances**: Forming strategic alliances with suppliers, distributors, or technology providers. This shall be discussed in detail in the section on GTM model for petrochemical marketing.

By effectively addressing these aspects, petrochemical companies can successfully navigate the complexities of the market and build strong, lasting relationships with their customers.

4.3 CHALLENGES FOR PETROCHEMICAL MARKETING AND TACKLING THEM

Discussed below are some key challenges for the petrochemical marketers to tackle.

1. **Price Volatility**
   a. **Flexible Pricing Models**: Developing flexible pricing models that can adapt to fluctuations in raw material costs and market conditions. In a

later section one such pricing mechanism (dynamic pricing) shall be discussed in detail.

   b. **Hedging Strategies**: Using financial instruments to hedge against price volatility and manage risk. This helps in not only improving the bottom-line but also mitigating supply chain management risks.

2. **Innovation and Differentiation**
   a. **R&D Investment**: Investing in research and development to create innovative products that meet evolving market needs. This helps tackle the dynamic business environment where ever-growing customer needs are met only with a consistent R&D effort.
   b. **Differentiation**: Differentiating products based on quality, sustainability, and technical support rather than just price. Some major bases of differentiation have been discussed later in the chapter.
   c. **Product Development**: Investing in R&D to develop new products that meet emerging needs. As end-users look for innovation and cost-optimization, R&D by petrochemical companies is important to come up to the expectations of their customers. Moreover, consistent R&D is also required to meet the ever-evolving standards of sustainability.
   d. **Process Improvements**: Enhancing production processes to reduce costs and improve efficiency. By this, the marketers may also be able to pass on some benefits to their customers.
   e. **Undertaking Sustainability Initiatives**: Petrochemical marketers are always expected to be developing sustainable products/services and sustainable practices so as to meet regulatory and consumer demands.

As discussed, marketing petrochemicals in a B2B context requires a nuanced approach that combines general B2B Marketing principles with the particular requirements of the petrochemical industry. Hence, by focusing on technical expertise, compliance on regulations, customer relationship management, and tailored marketing strategies, the petrochemical companies/marketers can effectively reach and engage their target markets.

4.4 SEGMENTATION, TARGETING, AND POSITIONING (STP) FOR PETROCHEMICAL MARKETING

The **Segmentation, targeting, and positioning (STP) model** is a strategic marketing framework that helps marketers identify, access, and reach their ideal customers in the best possible way. Thus, given the diverse range of products, applications, and markets involved in the context of petrochemical marketing, the STP model can be applied effectively. We discuss below, as to how STP can be applied in petrochemical marketing:

4.4.1 Segmentation

Segmentation involves apportioning the broad market into distinct groups of customers with similar requirements, expectations, behaviors, or characteristics. For petrochemical companies, the market segmentation can be done on the basis of several criteria as follows:

- **Product-Based Segmentation**: Dividing the market based on the type of petrochemical product, such as olefins (ethylene, propylene), aromatics (benzene, toluene), polymers (polyethylene, polypropylene), etc.
- **Industry-Based Segmentation**: Categorizing customers based on the industries they belong to, such as automotive, construction, pharmaceuticals, agriculture, and packaging.
- **Geographic Segmentation**: Segmenting the market based on regions or countries, considering local demand, regulations, and competition.
- **Customer Size and Value**: Segmenting based on the size and purchasing power of customers, such as large industrial manufacturers versus small enterprises.
- **Usage-Based Segmentation**: Dividing customers based on how they use the petrochemicals, for instance, raw materials for production, additives, or specialty chemicals.

4.4.2 Targeting

After identifying the market segments, the next step for petrochemical marketers is to choose which market segments to target. This involves evaluating the attractiveness of each segment and deciding where to focus marketing efforts. Key considerations include:

- **Market Potential**: Assessing the size, growth rate, and profitability of each segment.
- **Competitive Landscape**: Analyzing the level of competition and the company's ability to compete effectively.
- **Strategic Fit**: Ensuring the target segment aligns with the company's strengths, resources, and strategic objectives.
- **Customer Needs**: Understanding the specific needs, preferences, and purchasing behaviors of each segment.

In petrochemicals, a company might target segments such as large automotive manufacturers for specialty polymers or regional construction companies for basic chemicals.

4.4.3 Positioning

Positioning involves defining how the company's products are perceived in the minds of the target customers, relative to competitors. Effective positioning requires clear communication of the product's unique benefits and value propositions. Given the unique nature of products, for petrochemicals, positioning strategies might include:

- **Quality and Performance**: Emphasizing the superior quality, reliability, and performance of the petrochemicals.
- **Innovation and Technology**: Highlighting advancements in product formulations, sustainability, or production processes.
- **Cost-Effectiveness**: Positioning products as cost-effective solutions that provide excellent value for money.
- **Sustainability**: Focusing on eco-friendly products and sustainable practices to appeal to environmentally conscious customers.
- **Customization and Service**: Promoting the ability to offer tailored solutions and excellent customer support.

4.5 PETROCHEMICAL MARKETING MIX

This section briefly discusses the context of marketing mix framework for the petrochemical products. Thus, we start with the discussion on pricing for petrochemical products (including a sub-section on technology-enabled dynamic pricing model), followed by sales and distribution, integrating marketing communications, solutions-based marketing, GTM, and digital marketing.

4.5.1 Product

In the interest of brevity, the product-mix element of marketing mix is not discussed here as Chapter 1 provides a comprehensive information on various petrochemical products.

4.5.2 Price

Pricing petrochemical products involves a complex interplay of factors due to the volatile nature of raw materials, global market dynamics, and the specific needs of diverse industries. Here are key concepts and strategies involved in pricing petrochemical products:

1. **Cost-Based Pricing**
   This method involves calculating the total cost of production (including raw materials, labor, overhead, etc.) and adding a mark-up to ensure a profit. This is common in commodities where competition and cost structures are relatively similar.
2. **Market-Based Pricing**
   This requires a petrochemical marketer to look outwards to take appropriate pricing decisions. The two major pricing techniques are as follows:

a. **Competitive Pricing:** This involves setting prices that are based on the prevailing market rates and also compared with competitors' pricing strategies. Petrochemical companies often monitor competitor prices closely and accordingly fine-tune their own prices to stay competitive.
b. **Demand-Based Pricing:** This pricing method is about adjusting prices based on the demand for the petrochemical product. A higher demand can lead to higher prices, while conversely, a lower demand might require price reductions.

3. **Dynamic Pricing**
   a. **Spot-Pricing:** This technique involves pricing that fluctuates based on real-time market conditions. Spot prices are influenced by prevailing supply and demand factors and are commonly used for short-term transactions. Hence, it is called (on the) spot-pricing.
   b. **Contract Pricing:** Contrary to spot-pricing, this technique involves long-term contracts with fixed or variable pricing mechanisms. These contracts often include clauses that allow price variations based on changes in raw material costs, other input costs, and changes in taxes/ duties or to incorporate monetary impacts of any other economic indicators.
4. **Value-Based Pricing**
   This mechanism involves determining prices based on the perceived value of the product to the customer. That's because, based on their requirements, different customers have differing perceptions about values that they expect from various products/services. This includes understanding how much customers are willing to pay (WTP) for precise benefits such as higher quality, reliability, advanced features, etc.
5. **Additional Pricing Methods Contextual to Petrochemical Products**
   Apart from the ones discussed above, some more pricing techniques contextual to petrochemical marketing are discussed below:
   a. **Differential Pricing**
      i. Geographic Pricing Method: This entails adjusting of prices based on regional market conditions, accounting for variations in demand, competition, and different regulatory environments.
      ii. Customer-Based Pricing Method: Offering different prices to different customers based on their purchase volume, loyalty, or long-term relationship with the company.
   b. **Hedging and Risk Management**
      i. Hedging: Price hedging involves application of financial instruments like futures and options (F&O) to mitigate the risk of price volatility in raw materials. This also helps in stabilization of costs and ensures avoiding variability in pricing.
      ii. Forward Contracts: Agreements to sell products at a predetermined price at a future date, protecting against adverse price movements.

c. **Pricing Formulas**

Petrochemical marketers rely on commonly used pricing formulas for determining prices of petrochemical products. Some of these are as follows:

i. Feedstock-Based Pricing: This involves linking the price of petrochemical products to the cost of feedstocks (i.e. crude oil, natural gas). This makes sure that price adjustments are tantamount to changes in raw material (input) costs.

ii. Index-Based Pricing: This formula entails linking prices to established indices like Platts [2] and ICIS (International Commodity Intelligence Services) [3] that reflect market conditions and trends. Platts and ICIS indices are some of the most authentic indices available in the market.

6. **Factors Influencing Petrochemical Pricing**

Petrochemical pricing is incumbent of various factors. Some of them are described below:

a. **Raw Material Costs**: The variations in the prices of inputs like crude oil, natural gas, and other feedstocks directly impinge upon the production costs and, consequently, the pricing of petrochemical products.

b. **Forces of Supply and Demand**: These forces explain the market balance between supply and demand (which) influences prices. It is self-intuitive that (all other things remaining the same) an oversupply can drive prices down, while high demand or supply constraints have the tendency to push prices up.

c. **Economic Conditions**: Given the nature of petrochemical industry, major global economic trends, like GDP growth, industrial production, consumer spending, etc. affect demand for petrochemicals and subsequently the pricing.

d. **Regulatory Environment**: Prevailing environmental regulations, trade policies, and tariffs can always impact costs and pricing strategies for petrochemical marketers.

e. **Technological Advances**: The petrochemical industry is a dynamic one. And hence innovations in production processes or the development of new products have the propensity to affect costs and also the competitive dynamics.

f. **Cyclicality**: Given the cyclical nature of the industry, many petrochemical products may experience seasonal demand variations, which may result in influencing the petrochemical marketers' pricing strategies.

g. **Transportation and Logistics**: Petrochemicals products' marketing involves significant movements. Thus, the costs (and cost variations) associated with shipping, storage, and distribution also play a vital role in pricing, particularly for international markets.

7. **Pricing Models and Tools**

a. **Pricing Software**: Advanced pricing software helps companies analyze market data, forecast demand, and set optimal prices.

b. **Data Analytics**: Using big data and analytics to understand market trends, customer behavior, and competitive actions to make informed pricing decisions.
c. **Scenario Analysis**: Modeling different market scenarios to anticipate potential price changes and adjust strategies accordingly.

As discussed in detail above, effective pricing of petrochemical products requires an ample understanding of cost structures, market dynamics, customer value perception, and external factors influencing the petrochemical market. However, by leveraging a mix of cost-based, market-based, and value-based pricing strategies, petrochemical companies can manage the complexities of the market and attain sustainable profitability. We discuss below a case of one such pricing strategy for petrochemical products.

4.5.2.1 Dynamic Pricing for Petrochemical Products

As per a report by a consultancy firm [4], pricing excellence can increase profitability and the top line by 2 to 5 percent. The complete set of price drivers for petrochemical commodities must be identified, including those that are market-specific, customer-specific (e.g., strategic importance, cost-to-serve, value-added services, and alternatives available), internal (volume target, inventory position), and order-specific (size, urgency, and customizations). In order to arrive at customer-specific target pricing, players must shift from subjective *intuition-based* calls to more analytics-driven dynamic pricing approaches [5]. To ensure successful execution, the pricing process must be backed by strong processes and alignment of the main areas of responsibility for the sales teams. Establishing clear prices and margins for stock-keeping units (SKUs) and a value-based pricing strategy that is dynamic in nature are essential for petrochemicals in order to maximize value. By automating target price computations and assisting sales teams in maintaining the advantages, analytical and digital tools can be of great assistance. See Figure 4.2.

FIGURE 4.2 Dynamic Pricing—An illustration.

4.5.2.2 The Role of Digital Technology in Dynamic Pricing: A Case Study

The current era has witnessed an increased role of digital technologies in optimizing the business processes of the petrochemical industry. The industry seeks technological prowess to implement various measures to improve its profitability. Pricing is also one area where petrochemical marketers regularly seek expertise. The below section describes the case of a digital technology company, Tech Mahindra (www.techmahindra.com) offering its technological expertise [6] to implement the concept of dynamic pricing for the petrochemical industry.

As already discussed earlier, the product line of the petrochemicals sector range from specialty petrochemicals with unique end-uses to highly commoditized products, which are feedstock for plastic manufacturing plants. Petrochemical businesses, for their wide range of products, currently use the traditional pricing system, which decides a price for a certain product for a predetermined amount of time. Because the traditional pricing mechanism uses a formula-driven pricing based on global prices, freight, packing costs, and marketing expenses, there are good chances that this may not be the best pricing mechanism and value may be lost in the process.

This mechanism does not optimize the sales volume and does not necessarily take into account the nature of the product, the customer's willingness-to-pay (WTP), the customer's price-sensitivity and elasticity, or the ability of competitors to supply.

In contrast, *dynamic pricing* adopts a data-driven approach to pricing by utilizing data analytics and other modern digital technologies. It evaluates past data from all of the customer's purchases to provide insights into their price-sensitivity and willingness-to-pay. It then suggests a price that will optimize profits for the petrochemical businesses. It also considers every external factor that could influence prices, including supply and demand curves, seasonality, rivals' prices, disruptions, capacity expansions, etc. By taking into consideration all of the aforementioned factors and utilizing cutting-edge analytical tools, businesses can improve pricing performance to a higher degree of precision, which raises earnings and market share [6].

Petrochemical marketers are able to leverage and profit from the value-added by their product(s) by their end-user customers thanks to dynamic pricing. Given the nature of volatility in international crude oil prices, regular price adjustments are essential for petrochemicals products. For these kinds of products, margin maximization is made possible via dynamic pricing.

Implementing Dynamic Pricing

For companies to fully benefit from dynamic pricing, they need to adapt their technology, procedures, and manpower [7]. Given the scope of this section, the technological aspect is being discussed here.

Technology's role: To apply dynamic pricing, petrochemical marketers first need to set up the required infrastructure, such as a data platform with sophisticated analytics capabilities. This comprises both hardware and software that can extract and manipulate data to produce insightful outputs after ingesting it from a variety of internal and external sources. The following features need to be included in the platform:

1. Data Acquisition
2. Leveraging Industry 4.0 capabilities
3. Information Management
4. Statistical Visualization

Incidentally dynamic pricing doesn't give the intended benefits if the infrastructure lacks the necessary facilities. Following the setup of the technology and infrastructure and the migration of all pertinent data to it, the data scientists need to go through the following four steps [8] in order to derive maximum benefit of dynamic pricing mechanism:

1. **Exploratory Data Analysis:** The data team needs to compile all of the historical data for every customer and transaction on this data platform. Additionally, they need to obtain external data from this platform, such as supply-curves, competitors' prices, market prices, and trade discounts. It is necessary to validate and clean up all of the aggregated data. The data is analyzed next and explored to find trends and patterns.
2. **Clustering:** Given the vast number of consumers, the data scientist identifies a group of customers based on purchase volume or trend. Ideally, this involves analyzing each customer's transaction data. Clustering is the process of putting consumers together based on shared characteristics. To facilitate further action and analysis, clustering groups the clients who exhibit similar behaviors in reaction to a pricing choice. In order to perform clustering, a data scientist first aggregates the data at the customer-product level and chooses an appropriate time frame for calculating variables like:
   a. **Sales Volume:** Total volume of concerned petrochemical product bought during the period.
   b. **Price Elasticity:** Relationship between price variation and sales volume.
   c. **Customer Profitability:** Unit profit per customer (helps in identifying profitable customers).
   d. **Region Wise Freights:** Freight charges across regions (aids in route optimization).
   e. **Consignment Transportation Analysis:** Frequency of loading of consignments.
3. **Demand Curve Analysis:** To determine the appropriate pricing for the quantity that a client is willing to purchase, data scientists ideally should create a demand curve for each consumer at this stage. It should be constructed based on examining past purchasing patterns of customers, the state of the market, demand, and other pertinent factors. However, since that (for each customer) isn't feasible, an estimate of the demand curve is

made for every "cluster." Petrochemical firms are able to achieve "price-personalization"—that is, charging both price-sensitive and higher-paying consumers appropriately—by forecasting demand curves. This allows the company to maximize both volume and profit.

4. **Optimization:** Predicting the right pricing for every consumer is the ultimate aim of the previously mentioned steps. An optimization model is run to determine the best *price and volume* combination for each customer cluster once the demand curves for each customer cluster have been established. Put differently, optimization allows petrochemical businesses to assign a revenue ranking to each of their clients. Beyond maximizing the objective function—in this case, the overall profit margin—is what the optimization step considers. As inputs to the optimizer, businesses must additionally establish limitations like volume targets, inventory constraints, floor/cap premiums, ranking of customers, etc. The optimization algorithm looks for a single, optimal price for every consumer within the boundaries specified by the constraints.

Petrochemical marketers ought to implement dynamic pricing for their products in order to combat the mounting strain on margins. By passing on the higher costs to clients and charging more for products that are significantly different from the competition, they are able to raise their margins and tackle the volatility of feedstock prices [6]. These days, advanced analytics capabilities are easily accessible and are helpful in determining customers' willingness to pay as well as in estimating a value-based price for the thousands of small and medium-sized product combinations among customers. This optimizes returns and arrests value attrition.

Having discussed the technological aspect, for dynamic pricing to work it is equally important to have the support of leadership [5] to ensure that the current processes are adjusted to fit the new modes of operation. In addition, cross-hierarchy training is also desirable to ensure that the dynamic pricing method gives best results to petrochemical marketers.

4.6 SALES AND DISTRIBUTION MANAGEMENT

Sales and distribution management for petrochemical products involves efficiently getting products from manufacturing facilities to end customers while maximizing sales revenue and maintaining strong customer relationships. We briefly discuss the important aspects towards this:

1. **Sales Strategy**

   Designing an effective sales strategy involves the following steps:

   a. **Market Segmentation**: As also briefly discussed in a previous section, this process involves identifying the target markets and customers that are based on parameters like industry, geography, product type, buying behavior, etc.

b. **Sales Forecasting**: Using historical data, market trends, and customer insights to forecast sales volumes and plan production accordingly.
c. **Customer Relationship Management (CRM)**: Implementing CRM systems to track customer interactions, manage leads, and provide personalized service.
d. **Salesforce Management**: Recruiting, training, and managing sales teams to effectively communicate product value propositions and close deals.

2. **Distribution Channels**
   Identifying appropriate channels of distribution is vital to reach out to customers. Some of the types of these channels are discussed below:
   a. **Direct Selling**: This means selling directly to large industrial (B2B) customers. This is done often through dedicated sales teams or key account managers, who cater to these specific customers.
   b. **Distributors and Agents**: Partnering with distributors and agents to reach smaller customers or penetrate new markets.
   c. **E-Commerce Platforms**: Utilizing online platforms and portals to facilitate sales transactions and reach a broader audience.
   d. **Global Logistics**: Establishing efficient logistics networks for transporting products across regions and continents, including storage, shipping, and customs clearance.
3. **Pricing and Contract Negotiation**
   Given the nature and typicality of petrochemical marketing, pricing, and commercial issues are very crucial and should be handled diligently.
   a. **Price Negotiation**: Petrochemical market is at times susceptible to prices. Thus, negotiating pricing terms with customers, which are based on volume, contract-tenure, market conditions, etc. is crucial for the markers.
   b. **Contract Management**: Managing long-term supply contracts, including pricing adjustments, delivery schedules, and terms and conditions.
   c. **Risk Management**: Mitigating risks associated with price volatility, currency fluctuations, and geopolitical factors through hedging and contractual arrangements.
4. **Customer Service and Support**
   Customer service (both pre and post sales) is an intrinsic part of petrochemical marketing. This entails:
   a. **Technical Support**: Providing technical assistance to customers for product selection, application optimization, and troubleshooting is an important task for the marketers.
   b. **After-Sales Service**: This process involves offering post-sales support, including warranty management, product recalls, and maintenance services.
   c. **Complaint Resolution**: In order to minimize customer dissonance, resolving customer complaints promptly and effectively is very important. This also helps in maintaining satisfaction and loyalty.

5. **Market Development**
   Market development activity is aimed at the objective of ensuring market growth in all directions. Some of the important activities undertaken towards this task are as follows:
   a. **Market Research**: Analytical in nature, this entails conducting market research to identify emerging trends (existing, latent, and emerging), customer needs, and also competition-dynamics.
   b. **Product Development**: Collaborating with R&D teams to develop new products or formulations tailored to specific market segments.
   c. **Marketing Campaigns**: Launching targeted marketing campaigns to promote products, generate leads, and increase brand awareness.
6. **Regulatory Compliance**
   As also discussed in a separate context in the previous sections, petrochemical marketer needs to ensure compliance with all applicable regulations pertaining to the petrochemical industry. This can be achieved through:
   a. **Quality Assurance**: This activity involves compliance with industry standards and regulatory requirements for product quality, safety, and environmental protection (including reducing carbon footprint).
   b. **Export Controls**: Adhering to export regulations and trade restrictions imposed by international agreements and government agencies.
   c. **Documentation and Certification**: Obtaining necessary documentation and certifications,e.g., MSDS (Material Safety Data Sheet) [9], REACH [1] for product registration and clearance.
7. **Performance Measurement and Analysis**
   Taking stock of performance and initiating corrective measures wherever applicable is desirable from the petrochemical marketers. This involves the following:
   a. **Key Performance Indicators (KPI)**: KPIs, among others include quantitative factors like sales performance metrics such as revenue, volumes sold, market share (value/percentage/volumes), customer satisfaction, scores, etc.
   b. **Sales Analytics**: Analyzing sales data to identify trends, opportunities, and areas for improvement in sales and distribution processes.
   c. **Continuous Improvement**: Implementing strategies to optimize sales and distribution operations based on data-driven insights and feedback from customers and stakeholders.

Effective sales and distribution management is vital for petrochemical marketers to maximize market penetration, optimize revenue generation, and also maintain robust customer relationships. Petrochemical marketers, by implementing rigorous sales strategies, effective distribution channels, providing a customer-centric service, and effective performance measurement mechanisms, can augment their competitiveness and subsequently achieve sustainable growth in competitive markets.

In the contemporary era, the sales and distribution function for petrochemical business has also evolved with digitalization. Appropriate application of technology in sales and distribution decisions has the potential to bring about several benefits for petrochemical marketers. The upcoming section highlights how digitalization is being leveraged not only for routine decisions but also for models like GTM.

4.6.1 Digital Approach to Sales and Distribution Management in Petrochemical Business

Digitizing the lead-to-cash cycle is referred to as a digital approach to sales and distribution. Through the integration of ERP systems, business processes, and support services, digitization facilitates order acquisition, expands the sales funnel, and supports fulfillment. Petrochemical marketers may compete in online marketplaces regardless of the state of technology or the sophistication of sales procedures by utilizing customizable order management and fulfillment systems.

Digital applications provide sales teams and channel partners with a consistent interface to manage products, pricing, and promotions by providing real-time visibility into consumer data, transaction records, demand projections, and inventory levels at various locations. Additionally, analytical dashboards (See Figure 4.3) speed up the creation of quotes and orders as well as sales response times. Further increasing the productivity and efficiency of the sales force are built-in capabilities for commission calculations, order tracking, partner on-boarding, reporting, and order tracking.

There are tools that use machine learning and artificial intelligence to automate performance management, revenue, and marketing processes. Data from many

FIGURE 4.3 Analytical Dashboard.

sources is harmonized and easily integrated into an advanced ecosystem. By repurposing warehouse space for e-commerce fulfillment, rerouting product to alternate destinations, or personalizing marketing activities, it enables organizations with digital capabilities to meet internal limits, external difficulties, and emergent opportunities.

In addition to bolstering the distributor network, digital tools facilitate the execution of a direct-to-consumer approach. Customers may find dealers and distributors, browse product catalogs, and enable provenance monitoring with ease thanks to dashboard solutions that come with chat-bots and sophisticated search features. Digital technologies enable self-service for customers, reducing cost-to-serve.

Having understood the aspects of sales and distribution management for petrochemical products and application of digital tools in the same, the upcoming section deals with a unique GTM model for petrochemical marketing with the help of an applied illustration.

4.6.2 Evolving Go-to-Market (GTM) Models for Petrochemical Marketers

In marketing, a GTM strategy is a plan that outlines how a company will launch a product or service and reach its target customers effectively to achieve a competitive advantage [10]. It encompasses the actions required to deliver a product or service to the end customer, including positioning, sales strategy, pricing, and distribution. The key components of a GTM strategy include target market, value proposition, positioning, sales and distribution, channels, pricing strategy, marketing and promotion, sales strategy, customer support, metrics and KPIs, and feedback loop. A well-crafted GTM strategy helps ensure that the product or service reaches the right audience, meets their needs, and achieves business objectives efficiently.

Go-to-market (GTM) approaches are therefore necessary for petrochemical marketers to combat the commoditization of petrochemical products. GTM models require innovation or uniqueness, and the majority of channel partners and other intermediaries must provide significant value that broadens the client base and boosts customer interaction. Even if the petrochemical business may have benefited much from current GTM models up to this point, a substantial market evolution driven by some fundamental shifts is always anticipated.

1. In order to combat margin pressure, commodity sales must begin to shift toward low-overhead, straightforward, digital-first models. Commodity product margins are always under growing pressure as competition grows. Petrochemical marketers around the world are under increasing pressure to adapt to changing market conditions by embracing and experimenting with digital solutions. For instance, product-and application-specific online portals can function as a single point of contact for all inquiries on products and their uses, as well as for ordering, tracking orders,

and resolving complaints. To develop a truly minimal overhead model, this requires the assistance of digitalized supply chains as well as standardization of operating processes and business regulations (such as lead times, credit periods, and minimum order quantities). With the introduction of the digitally first brand *Xiameter* in the early 2000s to appeal to consumers who were price conscious for commodity silicone products, Dow Corning (now a part of Dow Chemicals; www.dow.com) was a pioneer in successfully demonstrating the success of this model [11]. When compared to a conventional Dow Corning business unit, many fewer employees were needed due to the web-enabled platform's reduced requirement for human connection with clients. The company was able to retain strong margins and boost its online sales to 30% of total sales (about three times the industry average) by implementing a unique value proposition. Many petrochemical companies can use this as a model to shift a sizable percentage of their existing commodity product client base to direct sales, doing away with middlemen.

2. Petrochemical marketers must transition from being product marketers to solution providers (as also discussed later in the chapter) in order to promote distinction and customer loyalty in direct sales of niche items. Large volumes and expanding product portfolios call for new channel arrangements and sales and marketing competencies. Customers expect more from vendors as end-applications get more complex. Producers of petrochemicals need to make an investment in building trusting connections with their customers by providing specialized services like collaborative application development, product testing, and troubleshooting, or other offerings that have a big impact on customer loyalty and purchase decisions. To do this, the sales force must be reoriented toward value-based selling and robust internal technical capabilities must be developed. Key significant clients should be the emphasis of a strong GTM model in order to demonstrate the ROI.
3. For petrochemical manufacturers, channel partners are still an essential component of their overall GTM strategy. Nevertheless, the composition of the channel mix must adapt as the market and the distribution environment change. Since the conventional business model of intermediaries is rendered obsolete by the widespread adoption of digital technology, enhanced supply chain infrastructure, and expanded access to formal finance, channel partners must redefine their roles. Players in the petrochemical industry, for instance, might improve their market share by collaborating with distribution firms and setting up internal R&D facilities and technical services teams. These companies can offer a broad range of value-added services, including bulk breaking, repackaging, portfolio selling, mixing and formulation, technical assistance, and vendor-managed inventory. For petrochemical marketers, these can greatly simplify sales and marketing. In order to manage channel complexity while diversifying their product portfolio, growing volumes,

and meeting changing customer expectations, marketers need to reconsider their channel structure in the future and find the ideal balance between small and large channel partners. Given their affordable and adaptable business approach, a number of the present small and medium-sized channel partners are also crucial for effectively catering to the dispersed consumer base for petrochemical products. But as producers seek to distribute their expanding volumes in both domestic and international markets and cater to a diverse clientele that demands value-added services, the role of big distributors will also remain crucial.

Subsequent to the discussion on GTM, it is contextual to understand the aspect of solution-based marketing for petrochemical business. The below section explains how a focus on solution (rather than just products) enables petrochemical marketers to have a competitive edge [4] [12].

4.7 SOLUTIONS-BASED MARKETING FOR PETROCHEMICAL BUSINESS

Solutions-based marketing, also known as solutions marketing, is a strategy focused on addressing the specific needs and challenges of customers by offering tailored solutions rather than just promoting products or services. This approach entails understanding the customer's problems and then offering comprehensive solutions that address these problems.

Some important aspects of solutions-based marketing are as follows:

1. **Customer-Centric Approach**: Petrochemical marketers need to have a deep understanding of the unique needs, pain-points, and goals of the customer.
2. **Problem Solving**: It is important to identify and articulating the problems that the customer may be facing and what can a marketer do to offer solutions that effectively address these issues.
3. **Value Proposition**: Customers always look for value. Thus, it is imperative on the part of the petrochemical marketer to effectively present the value and benefits of the solution in a way that echoes with the customer's particular situation.
4. **Customization**: Instead of generalization, the petrochemical marketer should aim at customizing the marketing proposition and the solution itself to meet the specific requirements of diverse customer segments.
5. **Consultative Selling**: It's desirable that marketers engage in a dialogue with customers to appropriately understand their needs so that personalized offerings can be made.
6. **Long-term Relationships**: With realizing the tenets of customer relationship management, the petrochemical marketer should aim at building and nurturing long-term mutually beneficial relationships with their customers by providing them with valuable solutions and support on a consistent basis.

Solutions-based marketing contrasts with traditional product-centric marketing, which tends to focus on the features and benefits of a product or service without necessarily connecting these to the specific needs of the customer. By prioritizing the customer's perspective and delivering targeted solutions, businesses can enhance customer satisfaction, loyalty, and overall success. Petrochemical business is largely perceived as a commodities business, attempts to adopt the tenets of solutions marketing to get a competitive advantage. This shall be manifested in a case study of a prominent petrochemical company in the below section.

4.7.1 Solution Marketing in Petrochemical Business: A Case Study of BASF

Background

BASF SE (www.basf.com/global/en.html), one of the largest chemical producers in the world, traditionally focused on selling individual petrochemical products. However, with increasing competition and evolving customer expectations, BASF recognized the need to adopt a solutions-based marketing approach. See Figure 4.4.

FIGURE 4.4 BASF Implemented a Solutions-Based Marketing Approach.

Implementation of Solutions-Based Marketing

With the objective of meeting customer needs in a better way and driving growth, this strategy aimed to provide comprehensive solutions to customer problems rather than just selling products.

Simultaneous to conducting a structured market analysis, BASF is intensively engaged with its customers. The market research and customer insights involved:

1. **Engaging with Customers:** Conducting in-depth interactions with key customers across various industries to understand their challenges and needs [13].
2. **Market Analysis:** Analyzed market trends and identified emerging demands in sectors like automotive, construction, agriculture, and consumer goods among others.
   After this step, BASF focused on taking the engagement with the customers to the next step, collaborating with them to identify the solutions to their problems and thus develop solution-based offerings.

Development of Solutions

1. **Collaborative Innovation:** BASF partnered with customers to co-develop solutions tailored to their unique requirements. This included joint research and development projects and pilot programs [14].
2. **Creation Centers:** BASF encourages its customers to explore the world of BASF Performance Materials worldwide and dive deep into their materials. BASF designers, engineers, and simulation experts offer their support to the customers in their early product development phase—for better and faster innovations [15].
3. **Integrated Offerings**: By doing that the company developed system solutions combining multiple BASF products and services to address specific customer challenges. For example, in the automotive industry [14], BASF offered solutions for lightweight materials that combine polymers, coatings, and additives.

Marketing and Sales Strategy

Solutions-based marketing requires some unique approaches and BASF was no exception to this. The following are the components of this approach.

1. **Value Proposition:** The BASF *Verbund* (means association/ collaboration/ combine in German) is what, BASF believes makes innovative solutions for a sustainable future possible. It shifted the value proposition from product features to customer outcomes. Marketing materials highlighted how BASF's solutions could improve efficiency, reduce costs, and enhance product performance for customers [16].

2. **Consultative Selling:** This is an important feature of solution-based marketing approach. And thus BASF also expects the sales team to adopt a consultative approach, focusing on understanding customer problems and recommending integrated solutions rather than pushing individual products [17].

Implementation and support are the key to the success of this approach. BASF exhibits this in an inimitable way as the means of:

1. **Technical Support:** Provided comprehensive technical support to help customers implement BASF solutions, including onsite assistance, training, and troubleshooting [18].
2. **After-Sales Service:** Offered robust after-sales support to ensure customer satisfaction and gather feedback for continuous improvement.

The outcomes of all these steps of solution-based marketing were very fruitful for BASF. These are as follows:

Enhanced Customer Relationships by building stronger, more collaborative relationships [19] specifically with customers [14] by addressing their specific needs and challenges. This increased customer loyalty and retention through the delivery of tailored solutions, which in turn led to **Business Growth** by achieving significant sales growth in key markets by offering differentiated solutions that provided clear value to customers [20] and also opened **new revenue** streams by entering into long-term service agreements and collaborative innovation projects. BASF was able to strengthen its **Market Differentiation** as it differentiated BASF from competitors by positioning the company as a partner in problem solving rather than just a supplier of chemicals. Finally, this **enhanced brand perception** as an innovative and customer-centric organization.

The additional outcomes were in terms of **Operational Efficiency,** which improved internal collaboration between product development, marketing, and sales teams, leading to more cohesive and effective market strategies. It also streamlined the product portfolio by focusing on high-value solutions that meet specific customer needs.

Case Example: Lightweight Automotive Materials

In the automotive sector, BASF's solutions based marketing approach led to the development of a comprehensive lightweight materials package [21]. This package included:

- **Performance Polymers:** Co-create lightweight, high-strength performance polymers that reduce vehicle weight without compromising safety [22].
- **Innovative Coatings:** Coatings that improve durability and aesthetics while reducing environmental impact [23, 24].
- **Additives and Compounds:** Specialized additives that enhance material performance and processing efficiency.

By integrating these components, BASF provided automotive manufacturers with a solution that addressed the critical issue of weight reduction, contributing to better fuel efficiency and lower emissions.

With better-performing materials, cutting-edge technological solutions and services, a global presence, and professionals by their side at every turn, BASF accelerates innovation. According to BASF, a consumer becomes more than just a customer when they partner with them. They can jointly create a better future for the world and themselves since they are co-creators with BASF.

BASF's adoption of solutions-based marketing transformed its approach to customer engagement and market positioning. By focusing on delivering comprehensive solutions tailored to customer needs, BASF not only strengthened its customer relationships but also drove significant business growth and market differentiation. This case demonstrates the power of solutions-based marketing in the petrochemical industry, highlighting the importance of understanding customer problems and delivering integrated solutions that provide tangible value.

4.8 INTEGRATED MARKETING COMMUNICATIONS FOR PETROCHEMICAL BUSINESS

Integrated marketing communications that rely heavily on advertising and promotion in the petrochemical business are crucial for maintaining and expanding market presence. Given the industry's technical nature and the importance of B2B (business-to-business) relationships, the marketing communications tend to be more specialized compared to consumer-focused sectors. Here are some prevalent ways in which integrated marketing communications play out for petrochemical business:

4.8.1 Advertising

1. **Trade Publications and Journals**
   Petrochemical companies often advertise in industry-specific publications and journals where their target audience, such as chemical engineers, procurement managers, and industry analysts, frequently look for information.
   See Figure 4.5.
2. **Industry Websites and Online Platforms**
   Advertising on websites and platforms dedicated to the petrochemical industry can help companies reach a highly targeted audience.
3. **SEM and SEO**
   Optimizing websites for search engines and running PPC campaigns ensures visibility when potential clients search for relevant products and services.
4. **Technical Webinars and Online Conferences**
   Hosting or sponsoring webinars on technical topics can position a company as a thought leader and generate leads.

FIGURE 4.5 Advertising Petrochemical Products. (Author's X Account: https://x.com/aeon_mask/status/1863841624628793682?s=46)

4.8.2 Promotion

- **Trade Shows and Exhibitions:** Participating in industry trade shows and exhibitions (Figure 4.6) is a key strategy. These events provide opportunities to showcase products, meet potential clients, and network with industry professionals.
- **Sponsorships and Partnerships:** Sponsoring industry events (Figure 4.7), research initiatives, or educational programs help build brand recognition and aligns the company with important industry developments.
- **Content Marketing:** Creating and distributing technical white papers, case studies, and research reports can demonstrate expertise and attract potential clients looking for solutions [25].
- **Corporate Social Responsibility (CSR):** Engaging in CSR activities, such as environmental conservation projects or community development, can enhance a company's reputation (Figure 4.8) and appeal to stakeholders who value sustainability.
- **Direct Sales and Relationship Marketing:** Given the B2B nature, direct sales efforts and relationship building through personalized communication and long-term partnerships are crucial. Sales teams often work closely with clients to understand their needs and offer customized solutions.
- **Email Marketing:** Targeted email campaigns with industry updates, product information, and insights can keep the company top-of-mind for current and potential clients.

FIGURE 4.6 Exhibit at a Petrochemical Trade Fair. (Clicked by Author.)

FIGURE 4.7 Petrochemicals Conference. (Clicked by Author.)

FIGURE 4.8 CSR Award to a Petrochemical Company. (Clicked by Author.)

- **Innovative Product Launches:** Launching new products with extensive media coverage, demonstrations, and trial offers can attract attention and generate interest from potential clients [26, 27].
- **Digital Transformation and Data Analytics:** Leveraging digital tools and data analytics to provide insights and improve customer engagement. This can include personalized marketing efforts based on customer data and predictive analytics to anticipate market needs.
- **Social Media Engagement:** While not as prominent as in consumer sectors, having a presence on professional social networks like LinkedIn can help in networking, sharing company news, and engaging with industry trends.

- **Public Relations (PR):** Managing public perception through press releases, interviews, and media interactions to highlight company achievements, sustainability efforts, and technological advancements.

4.8.3 Key Considerations for IMC in Petrochemical Marketing

- **Technical Expertise**: Petrochemical clients value technical knowledge and reliability. Demonstrating expertise through detailed product specifications, technical support, and high-quality content is vital.
- **Regulatory Compliance**: Ensuring all promotional materials and activities comply with industry regulations and standards to maintain credibility and avoid legal issues.
- **Sustainability Focus**: Increasing emphasis on sustainable practices means promoting eco-friendly products and processes can be a significant competitive advantage.

As discussed, integrated marketing communications in the petrochemical industry that largely includes advertising and promotion hinges on technical, relationship-based, and highly targeted promotional strategies to reach its specialized audience effectively. Unlike the consumer goods business, where creativity can be manifested in different ways for maximum benefits, the petrochemicals marketer has limited options for using creativity. However, as discussed above, despite that constraint, an array of techniques including (new-age) content marketing and digital communications are leveraged by the petrochemical marketing communication professionals for ensuring maximum visibility with efficiency among all stakeholders.

After discussing the framework of marketing in context with petrochemical marketing, it would be interesting to discuss the role of branding (normally the forte of B2C marketers) in the marketing of petrochemicals.

4.9 ROLE OF BRANDING IN PETROCHEMICAL MARKETING

Branding in petrochemical marketing plays a crucial role in differentiating companies and their products in a highly competitive and commoditized market. The key significances and characteristics of branding in this sector are as follows:

1. **Differentiation**: Branding provides differentiation to the petrochemical marketers in the following way:
   a. **Commodity Market**: Petrochemicals are often seen as commodities with little differentiation in their technical characteristics. It is here that branding enables petrochemical companies to differentiate their products and services beyond basic specifications. By doing this, petrochemical markets are also able to look at premium pricing for some of their products.
   b. **Competitive Edge**: A strong brand can create a competitive advantage, making it easier for companies to stand out in a crowded marketplace.

2. **Trust and Reliability**
   a. **Provides Quality Assurance**: A well-established brand invariably denotes quality and reliability. Thus, customers are more likely to trust and purchase products from a brand that enjoys a strong reputation.
   b. **Helps build Long-Term Relationships**: Consistent branding helps build long-term relationships with customers based on trust and reliability.
3. **Value Addition**
   a. **Perceived Value**: Effective branding can enhance the perceived value of petrochemical products, allowing companies to potentially command premium prices.
   b. **Customer Loyalty**: Strong brands foster customer loyalty, reducing price-sensitivity and encouraging repeat business.
4. **Market Expansion**: Strong brands enable market growth and petrochemical companies can look for the following:
   a. **Growing beyond domestic markets**: A strong brand equity facilitates market expansion and penetration, especially in international markets for the petrochemical products, where brand recognition can facilitate overcoming a lot of entry barriers.
   b. **Diversification**: Branding facilitates petrochemical marketers to diversify their product lines and enter new market segments with an identifiable and reliable brand identity.
5. **Communication and Promotion**: Branding helps streamline marketing communications by:
   a. **Effective Messaging**: Branding can provide a framework for consistent and clear communication of the petrochemical company's values, mission, and perceived benefits of their products. Brands often channelize marketing communication efforts by preventing the scattering of marketing communications.
   b. **Impactful Marketing Campaigns**: Strong brands enable effective and consistent marketing campaigns, thus augmenting the overall marketing strategy.

Characteristics of Effective Branding in Petrochemical Marketing

In the below section we attempt to understand the features of a result-oriented branding strategy for petrochemical marketing. We shall discuss aspects like consistency, value proposition, visual identity, emotional connection, positioning, and innovation.

1. **Consistency**
   a. **Brand Identity**: Branding enables consistent usage of logos, trademarks, color schemes, and branding messages across all platforms and promotional materials. This leads to cost efficiency also.
   b. **Quality Standards**: Effective branding denotes consistent product quality and customer service standards to the target markets.

2. **Clear Value Proposition**
   a. **Unique Selling Propositions (USPs)**: Effective branding helps in clearly communicating the unique features and/or benefits that differentiate the petrochemical product(s) from those of competitors' product(s).
   b. **Sustainability and Innovation**: Good topical branding campaigns that emphasize on sustainable practices and innovative solutions can enhance the brand's value proposition.
3. **Strong Visual Identity**
   a. **Logo and Design**: Brand identities like a distinguishable and professional logo/trademark/punchline along with a consistent design language mirror the brand's values.
   b. **Packaging and Appearance**: Effective branding helps considerate packaging and overall visibility, which in turn reinforces the brand's image. It is particularly helpful for petrochemical marketing, which otherwise has to be marketed like commodities.
4. **Emotional Connect**
   a. **Storytelling**: Branding campaigns can be used for storytelling so as to to create an emotional connection with the customers. It also helps in consistently highlighting the brand's historical antecedents, values, and the quality promise.
   b. **Communicating Customer Experience**: Branding also enables petrochemical marketers to focus on relevant customer experiences to build a positive connection with the company and its products/services.
5. **Enabling a Strategic Positioning**
   a. **Brand Positioning**: Effective branding can help in clearly defining the brand positioning that also aligns with the target audience's needs and preferences.
   b. **Competitive Analysis**: Branding enables petrochemical marketers to undertake an analysis of competitors' counterpart brands. Without branding it would be less effective to engage in such exercises.
6. **Ensuring Adaptability and Innovation**
   a. **Market Trends**: An effective branding strategy helps petrochemical marketers in staying abreast of market trends. It also helps in evolving the brand to meet changing customer needs by tweaking certain brand positioning and also launching new brands/extending existing brands in new product categories.
   b. **Incorporating Technological Advancements**: A good branding strategy acts as an enabler towards incorporating new technological developments in products and services and thus maintaining a contemporary and relevant brand image.

4.10 BRAND MANAGEMENT: THE CASE OF RELIANCE INDUSTRIES LIMITED

An illustration of the petrochemical brand portfolio of India's leading petrochemicals company; *Reliance Industries Limited* (RIL; www.ril.com)

provides a glimpse of branding of petrochemical products. As per RIL—whose revenue is in the range of 120 billion USD [28], and is on the Fortune Global 500 list [29]—their expertise lies in developing products and markets from *"concept to fruition"* and beyond. RIL's constant focus on innovation has helped them to emerge as a trendsetter in various markets and be known worldwide for their whole range of products. RIL asserts that each of their brands is a natural extension of their philosophy of excellence and the brands are tuned to not only the needs but also the aspirations of customers. Table 4.1 gives a snapshot of RIL's petrochemical brands [30].

TABLE 4.1
RIL's Petrochemical Brand Portfolio

Petrochemical Brand Portfolio of Reliance Industries Limited

S.N.	Brand	Description
1.	**RELAB**	Linear Alkyl Benzene (LAB).
2.	**Recron AirTherm***	Blend of warmth and lightness for thermal wear.
3.	**Recron AirLite**	Feathery softness and loftiness for sleep solution (to be used in pillows/cushions).
4.	**HEXaREL** ***Live Smart, in Comfort***	New-age graphene for smart, intelligent fabrics.
5.	**HEXaREL Fibrefill**	Graphene-enhanced new-age smart fiber.
6.	**RElan Fabric 2.0**	Specialty new-age fabrics for textile industry.
7.	**Recron**	Staple fibers, filament yarn, texturized yarn; twisted / dyed yarn.
8.	**Recron IDY**	High tenacity, low shrinkage industrial yarns.
9.	**Recron SHT**	Super high strength (SHT) and low shrinkage fiber.
10.	**Recron FANCYY**	Value-added creations, distinct weaves and easy usability.
11.	**Recron Stretch**	Stretch yarns for comfortable fit and freedom of movement.
12.	**Recron Cotluck**	Cotton look, cotton-feel yarns.
13	**Recron Superblack**	Dope dyed black with high consistency in shade
14.	**Recron Super Micro**	Super fine fibers, soft feel and touch.
15.	**Recron Superdye**	Produced with modified polymer enabling it to be dyed with cationic dyes which are known for bright shades.
16.	**Recron Kooltex**	Moisture management fiber that is cool, dry and fresh.
17.	**Recron Fibrefill**	Anti-microbial fabric, keeps fabric fresh for long.
18.	**Recron 3D Conjugate**	Fibrefill with unique 3D spring-like crimps. Highly resilient sleep and comfort products.

(continued)

TABLE 4.1 (Continued)
RIL's Petrochemical Brand Portfolio

Petrochemical Brand Portfolio of Reliance Industries Limited

S.N.	Brand	Description
19.	**Recron 3S**	Acts as "secondary reinforcement" in concrete which arrests cracks, increases resistance to impact/abrasion and greatly improves quality of construction.
20.	**Recron Certified-*The Sleep Experts***	Quality certified sleep products.
21.	**Recron FeelFresh**	Anti-microbial fabric, keeps fabric fresher for long.
22.	**Recron Recrobulk**	Hi-bulk fibers for soft feel and warmth.
23.	**Recron Green Gold**	Eco-friendly fibers made from 100% post-consumer polyester waste.
24.	**Recron Recosilk**	Yarns for a silken shimmer and swathes of color.
25.	**Recron FS-*Everlasting Protection***	High performance inherently Fire-Retardant polyester.
26.	**Recron Exclk**	Yarns for naturally soft silken hand-feel.
27.	**Recron EcoD**	For eco-friendly dyeing process.
28.	**relpet**	Polyethylene Terephthalate (PET)

* *Brands from S.N. 2 to 28 are Polyester Brands*
Source: www.ril.com/about/products-brands

As we discuss below, the branding strategy of petrochemical products by RIL, India's leading petrochemical manufacturer, holds significant importance for several reasons:

1. **Providing Market Leadership and Recognition**: Branding of petrochemical products helps establish RIL's products in the market, thus not only avoiding the commodity trap but also differentiating them from the competitors. Strong branding thus enhances product recognition and customer loyalty, consolidating RIL's position as a market leader in the petrochemical industry [31].
2. **Conveying Quality Assurance**: By branding its products, RIL can always assure customers of the quality and consistency of its products and services. This is crucial in industries where reliability and performance are crucial for B2B customers particularly those, such as in automotive, packaging, and construction industries.
3. **Perceived Value Addition**: Branding lends perceived value in the minds of customers about RIL's products. It is observed that customers are often willing to pay a premium for branded products they associate with superior quality,

innovation, and trustworthiness. Brands enjoying good brand equity are supposed to do that for the petrochemical companies. And RIL's brands are no exception.

4. **Enables Market Expansion**: Effective branding aids in RIL's efforts to expand into new markets and segments [32]. Recognizable and respected brands can facilitate entry in to new international markets, thus expediting international growth and partnerships [33–35].
5. **Innovation and Differentiation**: Branding facilitates RIL to showcase its innovations and distinct product features. This differentiation is a must in a competitive market where technological advancements (also by effectively conveying through branding) can provide a significant edge to the company [36, 37].
6. **Ensuring Customer Engagement and Loyalty**: It is a fact that strong brands have the capability to foster customer engagement and build a long-term customer loyalty. RIL's petrochemical brands can help create a sense of community and trust among the users, thus leading to repeat purchases and also word-of-mouth referrals [38].
7. **Conveying Sustainability and Corporate Image**: Branding provides a platform for RIL to communicate its pledge to sustainability and corporate responsibility. This is increasingly important as consumers and other important stakeholders appreciate good ESG (environmental, social, and governance) practices [39].
8. **Facilitating Marketing and Communication**: It's a truism that branded products make marketing and communication efforts more effective. Consistent branding across various platforms helps strengthen the company's message and values. This, in turn, makes RIL's petrochemical marketers have an easy connection with their target market participants [40].
9. **Competitive Advantage**: A strong brand can be a significant competitive advantage. It can protect RIL's market share against new entrants and help the company navigate economic fluctuations and industry changes more resiliently [41].

As discussed above, branding of petrochemical products by Reliance Industries Limited is a strategic tool that enhances market presence, ensures quality perception, adds value, facilitates market expansion, promotes innovation, fosters customer loyalty, communicates sustainability, supports marketing efforts, and provides a competitive edge.

The role and significance of branding in marketing of petrochemicals is further explained by a case study on **IndianOil** (A Fortune 500 Company) [42] petrochemical brand PROPEL.

4.10.1 Petrochemicals Brand Management: The Case of PROPEL

Leading Indian hydrocarbon player and a Fortune 500 Company, Indian Oil Corporation Limited (IndianOil; https://iocl.com/) has effectively managed its petrochemical brand, PROPEL, to establish a strong presence in the market. The below section discusses some key strategies and approaches that IndianOil employs to manage and promote the PROPEL brand:

1. **Brand Positioning and Identity**
   a. **Distinct Identity**: PROPEL has been positioned as a premium brand in the petrochemical segment, with a strong focus on quality and reliability. The brand name PROPEL implies "propulsion" or "impetus" with a promise of growth to the customers' businesses.
   b. **Logo and Visuals**: The brand uses a distinctive logo and consistent visual identity across all marketing materials to ensure easy recognition and recall.

   See Figure 4.9.
2. **Product Range and Quality**
   a. **Diverse Product Portfolio**: For a wide range of applications across industrial, commercial, and domestic segments, IndianOil offers a full range of products under the umbrella brand PROPEL, covering all segments of petrochemicals, i.e., linear alkyl benzene (LAB), purified terephthalic acid (PTA), paraxylene (PX), mono ethylene glycol (MEG), polypropylene (PP), linear low-density polyethylene (LLDPE), high-density polyethylene (HDPE), etc.
   b. **Other Product Offerings under PROPEL**: Apart from petrochemicals, a range of other related product offerings and their applications under the brand PROPEL are as follows [43]:
      i. **Carbon:** IndianOil produces petcoke, carbon black feedstock (CBFS), and synthetic rubber to support industries engaged in production of tires, printing inks, plastic colorants, etc.
      ii. **Optical Films:** IndianOil also produces PTA and MEG, which are used in production of optical films such as bi-axial oriented polyester film featuring well-balanced characteristics to improve the visibility of touch panels.
      iii. **Food Packaging Industrial and Medical Films:** IndianOil has the capability to tweak characteristics and produce grades with added functionalities for specialty food and medical packaging, electronics parts, automobiles, and building materials.
      iv. **Fibers and Textiles:** With a world-class PX/PTA plant producing polyester intermediates, PROPEL grades can be used for a wide range of applications including active-wear, interiors, and industrial applications.
      v. **PP Composites and Functional Moldings:** PROPEL grades can be used to produce a variety of products such as aluminum composite materials, plastic film-laminated steel sheets, building materials, partitions, sports goods, industrial use plates, etc.

FIGURE 4.9 PROPEL Brand Petrochemical Product. (Clicked by Author.)

vi. **Agribusiness and Infrastructure Solutions:** PROPEL grades are suitable for use in producing agricultural materials such as drip laterals for irrigation pipes for submersible pumps, agri-nets for plant growing systems, etc. Contributing to a sustainable society. IndianOil petrochemicals can be used to produce piping materials, logistics materials and adhesive tapes, cable ducting, repair and reinforcement materials for roads and ridges, artificial turfs for boosting sports infrastructure, etc.

c. **Qualiy Assurance**: IndianOil emphasizes stringent quality control measures to ensure that PROPEL products meet international standards. This commitment to quality helps in building customer trust and loyalty. In order to provide high-quality items under the brand, PROPEL claims to have a cutting-edge R&D push. IndianOil has accelerated the development of new technologies in emerging fields, merged its numerous core technologies, and expanded its technological platforms in order to achieve this goal. Its tight collaboration throughout the Divisions enables it to react quickly to developments in the market. By basing its research and product application efforts on the pillars of comfort, health, and sustainability, IndianOil hopes to stand out from the competition and offer distinct solutions with distinct value propositions. It has a special R&D facility-one of Asia's best R&D centers—in Faridabad, India, close to the nation's capital.

With more than 800 active patents, it has been recognized for decades of ground-breaking work in pipeline transportation, nanotechnology, alternative fuels, biotechnology, petrochemical, and polymer (development of catalysts, niche grades, additives, etc.) as well as lubricant formulation. In addition to providing technical support in customer trials to improve product quality and lower production costs, its Petrochemicals and Polymers vertical conducts research in polyolefin catalyst, drag reducing agents, new grade development and post reactor modification, additives, troubleshooting and developing packaging materials, mainly polymeric, and benchmarking. IndianOil R&D, a developing Centre of Excellence in *Ziegler-Natta Catalysis* for polymer research, has generated new products and grades in response to market demands and assisted in standardizing its polymer grades. It makes use of general chemistry's fundamental technologies, such as material design, synthesis, and analysis, to create innovations that address both current and foreseeable technological needs [43].

3. **Market Expansion and Reach**
   a. **Wide Distribution Network**: PROPEL benefits from IndianOil's extensive distribution network, which ensures that products are readily available across India and in international markets.
   b. **Export Strategy**: The brand also targets international markets, leveraging INDIANOIL's global reach to expand its footprint and increase market

TABLE 4.2
The Global Footprint of PROPEL

Global Footprint of PROPEL	
Region	**Products**
North America	Polymers; DEG/TEG
South America	Polymers; DEG/TEG
Russia	Polymers
Central Asia	Polymers; DEG/TEG
Europe	Polymers; DEG/TEG
Africa	Polymers; LAB
Australia	LAB
China	Polymers; DEG/TEG
South East Asia	Polymers; DEG/TEG
South Asia	Polymers; DEG/TEG; LAB
Middle East	Polymers; LAB; Benzene
Japan	DEG/TEG
South Africa	LAB

DEG: Diethylene Glycol; TEG: Triethylene Glycol; LAB: Linear Alkaline Benzene

Source: https://iocl.com/pages/petrochemicals-overview

share. Apart from that, an extensive international presence enhances the overall brand image. Figure 4.10 gives an idea about PROPEL's global presence.

See Table 4.2.

4. **Customer Engagement and Support**
   a. **Technical Support**: IndianOil provides robust technical support to its customers, helping them optimize the use of PROPEL products in their applications. This includes technical consultations, troubleshooting, and after-sales service. IndianOil has set up two Product Application Development Centers (PDAC) in India, in Panipat and Paradip cities, in support of this strategy. The state-of-the-art PADC support innovative applications and specialized grade development driven by client needs, as well as commercial use. They also provide technical support to customers for product development and optimizations of polymer processing and quality control. By acting as a liaison between the company's end-user clients, marketing infrastructure, and polymer facilities, PADC creates novel grades, applications, and formulations that meet market demands. The Department of Scientific and Industrial Research (DSIR), Government of India's Ministry of

Science and Technology has recognized both PADC, Panipat and Paradip [43].

b. **Customer Feedback**: IndianOil actively seeks customer feedback for PROPEL to continuously improve its products and services, ensuring that they meet evolving customer needs.

5. **Sustainability and Innovation**

a. **Sustainable Practices**: IndianOil integrates sustainability into its operations and product development. PROPEL products are developed with a focus on minimizing environmental impact, which appeals to environmentally conscious customers.

b. **R&D and Innovation**: As discussed above, continuous investment in research and development ensures that PROPEL stays at the forefront of technological advancements in the petrochemical industry. This

FIGURE 4.10 A snapshot of PROPEL Print-Media and Social Media Advertisement. (Author's X Account: https://x.com/aeon_mask/status/1863831725425480127?s=46)

innovation drives the development of new products and enhancements to existing ones.

6. **Marketing and Promotion**
    a. **Integrated Marketing Campaigns**: PROPEL is promoted through a mix of traditional and digital marketing channels. This includes advertising in industry publications, participation in trade shows, and digital marketing strategies such as social media engagement and online advertising.
    b. **Brand Communication**: The brand's messaging emphasizes its core values of quality, reliability, and innovation. INDIANOIL uses storytelling to highlight the benefits and applications of PROPEL products, creating an emotional connection with customers.

Below figures give an idea about PROPEL's integrated marketing communication efforts, including brand communication to its various stakeholders.

See Figures 4.10–4.13.

7. **Strategic Partnerships and Collaborations**
    a. **Industry Collaborations**: IndianOil collaborates with various industry stakeholders, including suppliers, customers, and academic institutions, to foster innovation and stay ahead of market trends. And PROPEL is always at the cusp of leveraging these collaborative efforts. The PROPEL team participates in various events to benefit from the end-user and customer touchpoints that these events offer [44].
    b. **Joint Ventures**: Strategic partnerships and joint ventures help INDIANOIL enhance its capabilities and expand its product offerings under the PROPEL brand [43].
8. **Corporate Social Responsibility (CSR)**
    Community Engagement: IndianOil engages in CSR activities that benefit the communities in which it operates. This includes initiatives in education, healthcare, and environmental sustainability, which enhance the company's overall brand image.

FIGURE 4.11 One of the web banners of PROPEL. (Author's X Account: https://x.com/aeon_mask/status/1863821003463537050?s=46)

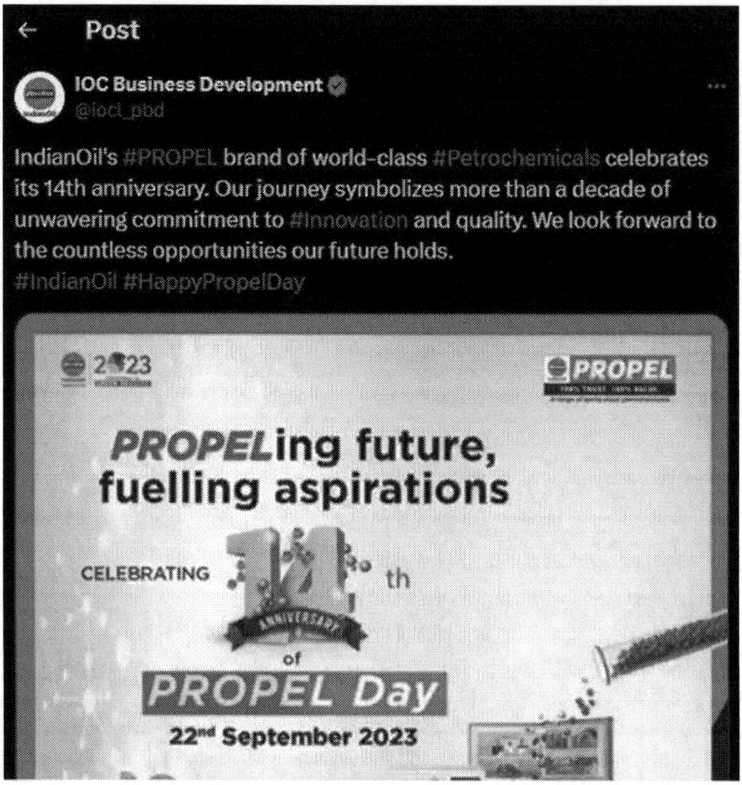

FIGURE 4.12 PROPEL Brand Communication at Social Media Platform X. (Author's X Account: https://x.com/aeon_mask/status/1863821485028278736?s=46)

As discussed, IndianOil manages its PROPEL brand through a combination of strategic positioning, a strong emphasis on quality, extensive distribution networks, customer engagement, sustainability initiatives, innovative marketing, and strategic partnerships. These efforts have helped establish PROPEL as a leading brand in the petrochemical industry, both in India and internationally.

In summary, branding in petrochemical marketing is essential for differentiation, building trust, adding value, and facilitating market expansion. Effective branding is characterized by consistency, a clear value proposition, strong visual identity, emotional connection, strategic positioning, and adaptability. These elements help petrochemical companies create a strong and recognizable presence in the market, ultimately leading to sustained growth and success.

After understanding the crucial and value-adding role of branding, the below section discusses how customer relationship management boosts the efforts of petrochemical marketers.

FIGURE 4.13 PROPEL Leverages YouTube for Its Integrated Marketing Communications Campaign. (Author's X Account: https://x.com/aeon_mask/status/1863820324820668529?s=46)

4.11 CUSTOMER RELATIONSHIP MANAGEMENT (CRM) IN PETROCHEMICAL MARKETING

Customer Relationship Management (CRM) is critically significant in the marketing of petrochemicals for several reasons. As the petrochemical industry is often characterized as having commoditized in nature, effective CRM practices have the potential to lend an edge to petrochemical marketers and thus effectively tackle the competition [45–48].

We discuss below the important points that describe the significance of CRM practices in petrochemical marketing:

1. **CRM's role in Building and Maintaining Long-Term Relationships**
   a. **Building Customer Loyalty**. CRM activities aid in building long-term mutually beneficial relationships between petrochemical marketers and customers, which is essential in an industry-often having commoditized characteristic-where customer loyalty can result in sustained business.
   b. **Facilitating Personalized Interaction**: Petrochemical marketers, by effectively utilizing the CRM mechanism, can leverage customer analytics, thus enabling them for personalized communications and services.

2. **Enhancing Customer Satisfaction**
    a. **Rendering Efficient Services**: Petrochemical marketers can utilize CRM systems to smoothen the process of dealing with customer queries and complaints, thus ensuring timely and effectual resolution of the same.
    b. **Extending Quality Assurance**: Marketers can leverage CRM systems' help in maintaining high quality and meeting customer expectations by utilizing the regular feedback collection and analysis of the same.
3. **Enhancing Sales and Marketing Effectiveness**
    a. **Targeted Marketing**: Contemporary CRM tools enable petrochemical marketers to segment their markets and subsequently target them with customized marketing campaigns. This helps in improving the effectiveness of these efforts.
    b. **Obtaining Sales Insights**: CRM systems deliver valuable data and insights into customer purchase patterns and trends, thus helping petrochemical sales teams to forecast demand and tweak their strategies accordingly.
4. **Augmenting Product Development and Innovation**
    a. **Providing Customer Feedback**: CRM processes enable the systematic collection and analysis of customer feedback for the petrochemical marketers, which subsequently can be utilized to provide further inputs to the product development and innovation teams.
    b. **Understanding Market Trends**: Knowing customer needs and preferences through CRM helps the petrochemical markets to stay ahead of market trends and thus develop products that not only meet emerging demands but also gives them a competitive advantage.
5. **CRM Increases Operational Efficiency**
    a. **Makes Processes Streamlined:** petrochemical marketers derive operational efficiencies from CRM systems as they help automate various processes such as order tracking, sales follow-ups, customer communication, etc.
    b. **Facilitates Data Integration**: CRM processes can be aligned with other simultaneous business processes. For example, integration of CRM with other systems like ERP and supply chain management can ensure a seamless flow of information across the organization, thus enhancing overall efficiency.
6. **Improving Customer Retention and Acquisition**
    a. **Customer Retention Measures**: CRM aids the petrochemical marketers in their understanding of customer behavior and needs. This enables them to implement effective retention strategies such as loyalty programs, customized offers, and proactive service for long-term mutual benefits.
    b. **Customer Acquisition Strategies**: Marketers can use CRM tools to identify potential leads and subsequently convert them into customers by tracking interactions (with the help of CRM processes) and ensuring follow-up at appropriate times.

7. **CRM Facilitates Better Decision Making**
    a. **Provides Data-Driven Insights**: Petrochemical marketers find CRM systems extremely useful as they provide a comprehensive data analytics suit, thus enabling them to make informed decisions based on real-time data.
    b. **Enables Performance Tracking**: It is important for petrochemical marketers to keep a tab on their marketing efforts. CRM processes can help them towards their efforts of tracking the performance of marketing campaigns, sales strategies, and customer service initiatives.
8. **CRM Enhances Functional Collaborations**
    a. **Cross-Departmental Collaborative Programs**: Marketers leverage CRM systems to aid better collaboration between various departments such as sales, marketing, and customer service. This enables an integrated approach to customer management.
    b. **Knowledge Sharing through CRM Processes**: CRM helps generate a significant amount of data. This data when put under a centralized customer information database, can allow for easy knowledge sharing and ensuring that all customer-facing employees have the information (on real-time basis or on times of their convenience) they need to provide exceptional service to their customers.
9. **CRM Supports Worldwide Operations**
    a. **Facilitates Scalability**: Petrochemical marketers find the utility of CRM systems—which are scalable by nature—and thus can support the complex needs of large petrochemical companies who operate in multiple regions. This brings about cost efficiency for them.
    b. **CRM facilitates Compliance and Documentation**: CRM systems also help the petrochemical marketers in maintaining detailed records and documentation. Such databases are crucial for compliance with requisite international trade regulations and standards.
10. **CRM aids in Strengthening Brand Image**
    a. **Consistent Communication with Customers**: Regular communication and customized touch with customers strengthens the brand image and reputation of a petrochemical company. CRM, by its very features, ensures consistent, cost-effective, and professional communication with customers.
    b. **CRM Ensures Positive Customer Experiences**: CRM helps in effectively managing customer relationships and thus enhances customer satisfaction. This helps the petrochemical marketer towards building a strong and positive brand image.

See Figure 4.14.

Having seen how CRM is an important tool in the arsenal of a petrochemical marketer, it would be now pertinent to know how several global petrochemical companies have successfully implemented customer relationship management

FIGURE 4.14 Customer Relationship Management.

(CRM) systems as part of their marketing strategies. We discuss below some noteworthy examples of some of these companies and how they leverage CRM for their strategic marketing efforts:

1. **ExxonMobil Chemical:** Global petrochemicals giant, ExxonMobil Chemical, uses a sophisticated CRM system to manage customer interactions and data. The company utilizes CRM for various marketing processes, including segmentation of customer base, which enables it to customize its marketing campaigns and ensure personalized communications. The usage of CRM processes also helps the company in monitoring customer feedback and complaints/queries, thus ensuring timely resolutions and improved overall customer satisfaction. Additionally, the company leverages CRM analytics to obtain useful insights into customer needs, which provides inputs for product development and innovation [49, 50].
2. **BASF:** BASF, a prominent global chemical company, has demonstrated fruitful integration of CRM into its sales and marketing processes. For example, BASF's CRM system captures comprehensive customer data across its various touchpoints, which is then used to create comprehensive customer profiles. The information obtained from these profiles is used to carry out targeted marketing efforts, such as customized promotions, loyalty programs, etc. The CRM system also facilitates effective coordination

between sales and marketing teams, thus leading to effective cross-selling and upselling strategies [51].

3. **Dow Chemical:** Global petrochemical giant, Dow Chemical, employs a robust CRM platform to augment customer engagement and thus improve operational efficiency. The CRM system at Dow Chemical also monitors customer interactions and preferences, thus enabling the marketing teams to offer personalized service and communications. CRM platform also supports the sales team with real-time data on customer behavior, purchase history, etc., and enables them towards strategic sales planning and execution. CRM analytics also helps Dow's marketing teams to identify market trends and customer needs, thus driving product innovation and competitive positioning [52, 53].
4. **Shell Chemicals:** Another prominent petrochemicals player, Shell Chemicals, also integrates its customer service and marketing operations with its CRM process. The CRM system at Shell Chemicals aids the company's objective of delivering outstanding customer service. It does it by capturing relevant customer interactions and feedback, which are then analyzed to improve service delivery and product offerings. The CRM system also enables Shell to run targeted marketing campaigns and monitor their effectiveness, and subsequently adjusting Shell's marketing team's strategies in real-time based on customer response [54].

Thus, after also understanding how some prominent global petrochemical companies leverage CRM for enhancing their marketing processes, we can summarize various benefits of deploying a structured CRM strategy as follows:

1. **CRM Facilitates Good Customer Understanding**
   Petrochemical companies use CRM processes to gain deeper insights into customer needs and preferences, enabling them to deploy more customized and result-oriented marketing strategies.
2. **CRM Helps in Better Customer Service**
   Petrochemical marketers find CRM systems very helpful in streamlining their customer service processes, thus ensuring timely responses and resolutions to customer inquiries and issues. This helps in enhancing customer satisfaction.
3. **Better Operational Efficiency**
   CRM helps in automation of sales, marketing, and service processes, which reduces manual efforts and errors. This, in turn, leads to greater efficiency and productivity for petrochemical companies.
4. **CRM aids in Data-Driven Decision Making**
   Data plays an important role in the business processes of petrochemical companies. CRM analytics provide valuable data on important aspects like customer behavior, market trends, etc., which helps in informed

decision-making and strategic planning for petrochemical marketers and other important functions.

5. **CRM Facilitates Stronger Customer Relationships**
One of the important outcomes of adopting CRM processes is that it generates detailed customer profiles and history. These can be used by petrochemical companies to foster stronger and more personalized relationships with their customers, thus augmenting loyalty and customer retention.
6. **CRM aids in Scalability and Global Reach**
One of the challenges petrochemical companies face is ensuring consistent customer relationship management across different regions and markets. CRM systems, with their prowess, help the companies towards meeting this challenge.

Thus to sum up, as discussed in the above sections, global petrochemical industry and their marketing functions have successfully deployed CRM systems to augment their marketing programs. The CRM systems, which are technology enabled, help the companies to better understand and serve their customers, streamline their various operations, and enable them to make data-driven decisions. All these eventually lead the companies and their marketing teams to improve customer satisfaction and achieve business excellence.

CRM is therefore crucial in the marketing of petrochemicals due to its role in building long-term relationships, enhancing customer satisfaction, improving sales and marketing effectiveness, driving product innovation, increasing operational efficiency, and supporting strategic decision making. Petrochemical companies and marketers, by leveraging CRM, petrochemical companies can gain an edge in a competitive market, retain and acquire customers more effectively, and eventually attain continued business success. After having gone through all relevant conceptual and applied aspects of marketing of petrochemicals, it would be appropriate to conclude the chapter by illustration of a case study of digital marketing in the petrochemical industry.

4.12 DIGITAL MARKETING IN THE PETROCHEMICAL INDUSTRY: THE CASE OF DOW CHEMICAL COMPANY

Digital marketing in the context of B2B (business-to-business) marketing-the realm in which petrochemical marketing also lies refers to the strategic use of digital channels to promote products or services to other businesses. This involves leveraging online platforms such as websites, social media, email, search engines, and content marketing to reach, engage, and convert business clients. Key tactics include SEO, PPC advertising, content creation, and data-driven campaigns aimed at building relationships, generating leads, and driving sales among a business-focused audience. The goal is to enhance visibility, attain market leadership, and facilitate informed purchasing decisions in a professional context.

Dow embarked on the journey of digital marketing with objectives of enhancing customer engagement, generating leads, and drive sales growth through a comprehensive digital marketing strategy. Digital is not new for businesses in the modern day. Providing clients with digital channels for interaction can be the difference between surviving and succeeding in the current challenging economic environment. For the 125-year-old Dow Chemical Company (Dow), one of the top producers of petrochemicals worldwide, this was also the case.

The requirement for businesses to deliver such digital solutions—regardless of audience—with increased speed and precision has shifted. As with B2C customers, even B2B clients of today need personalization. Additionally, there's a solid incentive for B2B businesses to invest in digital technology: And for businesses this was also not without a reason as after applying customization, 80% of businesses noticed an increase in sales [55].

When Dow began its digital transformation journey in 2020, they witnessed remarkable outcomes right away. They saw a 45–50% rise in revenue from digital sales by 2021, and by 2022, they were on pace to reach their target of $12 billion from digital sales.

Engaging with customers irrespective of their locations was one of Dow's guiding principles towards their digitalization journey. That basically entailed utilizing digital tools to comprehend their clients and identify their demands at every stage of the consumer journey.

Like a lot of well-known businesses, Dow was faced with a decision. Till ten years ago, companies would be considered easy to do business with, but customer expectations have changed due to digital disruption. Digital technology, such as fine-tuned search, was crucial in their sophisticated system because two out of every three visitors to www.dow.com involve using their search features [56, 57].

The Dow digital acceleration team started their digital transformation by identifying customer journey touchpoints through the development of a centralized customer data platform (CDP). They were able to learn a great deal about their clients as a result, including their identities, locations, languages spoken, preferred methods of conducting business, and the goods, markets, and application areas that pique their interest (based on their search patterns). Dow's search features include more sorting choices, icons to indicate format, and facets for rapid result refinement (Figure 4.15).

To obtain a very strong and cohesive picture of their clientele, this also required combining data from various traffic sources and touch points into the newly constructed digital platform, which includes the website, CRM, email, campaign management tools, call-centers, and more. For a firm with as diversified a customer base as Dow's, that was a tough order. It consists of decision-makers, influencers, scientists, and R&D divisions. Distributor partners and other internal and external stakeholders also use their digital platforms. The spoken languages and geographic locations of this incredibly varied population differed as well [58].

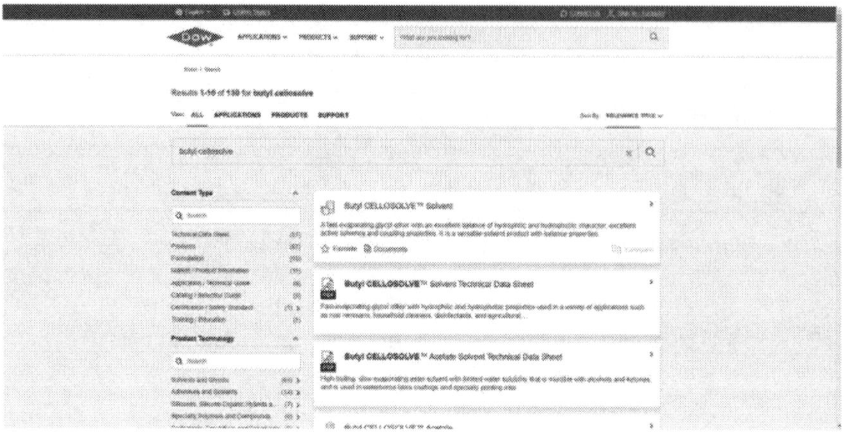

FIGURE 4.15 Dow's Search Refinement. (https://coveo.com/blog/dows-digital-acceleration-strategy/)

The cloud computing solution provided by the digital platform gave Dow's search a digital edge and made it considerably more intelligent. Based on real data from user interactions, the search becomes more intelligent with time. Ultimately, Dow's achieved its objective of developing a flawless search interface that makes it easy and quick for consumers to find what they were looking for.

One technique to maximize the user experience outside of the search box was to employ dynamic recommendations. A machine learning-driven AI recommender (Figure 4.16) charts user behavior over time. By making content recommendations based on the user's prior interactions, it produced more personalized experiences.

Priority was given to high-traffic touchpoints where they had the greatest interaction with their end-users. However, the website was only a single component of the whole digital transition.

With the philosophy that some personalization is preferable than none, Dow also gradually moved toward personalization using a tiered model.

One level of customization to many was known as tier one. Based on the visitor's onsite user behavior, Dow was able to learn a little bit about them. On the basis of that, content was curated.

Dow has redefined its marketing efforts through its digital transformation journey, showcasing how AI and data are the future frontiers that will drive corporate innovations. Additionally, this case explains how marketing professionals can improve their proficiency and efficiency by investing in AI-powered self-service platforms. By making these expenditures, Dow Chemical was able to differentiate itself from the competition and offer individualized customer experiences [28] [29] [30].

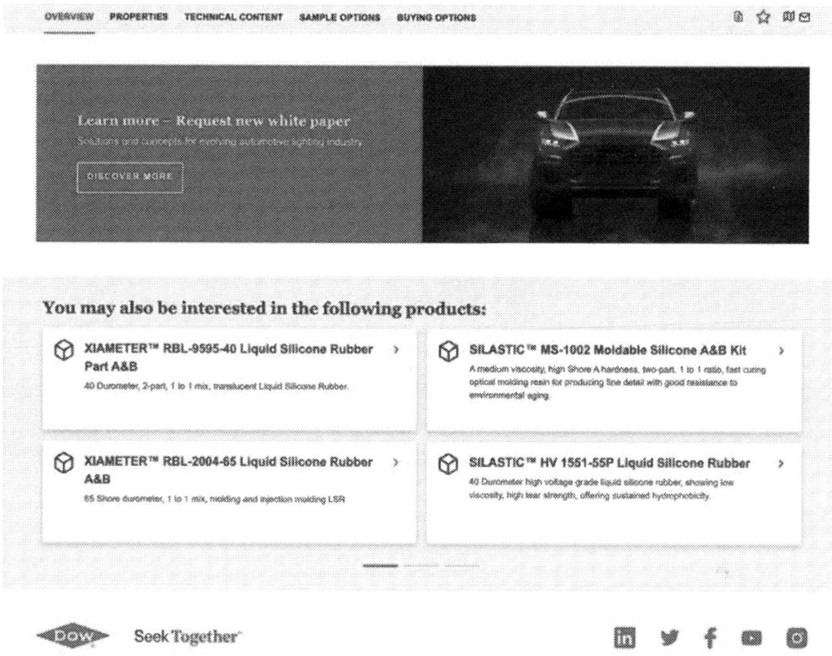

FIGURE 4.16 AI Based Recommender. (https://coveo.com/blog/dows-digital-acceleration-strategy/)

4.13 CONCLUSION

The chapter discussed various strategic and operational marketing aspects of the petrochemical industry in detail. As we saw, these aspects are multifaceted, involving strategic planning, market analysis, and innovative approaches to meet global petrochemical products' demand. Effective marketing not only augments brand visibility and competitiveness, but also promotes sustainable practices and productive customer relationships. Further, as the petrochemical industry evolves, leveraging digital technologies and data analytics becomes critical for adapting to emerging market trends and tackling regulatory changes. As mentioned in the chapter, ultimately, the market success of petrochemical companies depends on their ability to successfully steer these complexities and deliver value through efficient, environmentally responsible, and customer-centric strategies.

REFERENCES

1. European Chemical Agency. Understanding REACH. (n.d.). https://echa.europa.eu/regulations/reach/understanding-reach
2. S&P Global. (n.d.). Explore S&P Global. Retrieved May 24, 2024, from www.spglobal.com/commodityinsights/en

3. ICIS. Explore. (n.d.). Retrieved May 24, 2024, from www.icis.com/explore
4. Kearney. Riding India's Petrochemical Wave. (n.d.). May 24, 2024, from www.kearney.in/article/-/insights/riding-india-s-petrochemicals-wave
5. Burns, D., & Schottland, D. (2018, July 25). The formula for better pricing in chemicals. Bain & Company. www.bain.com/insights/the-formula-for-better-pricing-in-chemicals/
6. Nidhi. (2022). Dynamic pricing for petrochemicals. Tech Mahindra. https://files.techmahindra.com/static/img/pdf/dynamic-pricing-for-petrochemicals.pdf
7. Brégé, C., Langkamp, D., Beckett, M., Bourgouin, L., Chu, M., & Poirmeur, P. (2020, November 20). Debunking the myths of B2B dynamic pricing. Boston Consulting Group. www.bcg.com/publications/2020/dynamic-pricing-b2b-myths
8. Mishra, M., Theja, P., & Dasgupta, P. (2019, November 13). Transforming pricing for oil & gas wholesales: Personalized pricing for petroleum products. BCG Gamma. https://medium.com/bcggamma/transforming-pricing-for-oil-gas-wholesales-personalized-pricing-for-petroleum-products-a673f1796396
9. MSDS. Safety data sheet. (n.d.). Retrieved May 24, 2024 from https://chemicalsafety.com/sds-search/#:~:text=A%20Safety%20Data%20Sheet%20(SDS,Hazard%20Communication%20Standard%20(HCS)
10. Gartner. GTM. (n.d.). Retrieved May 23, 2024 from www.gartner.com/en/sales/glossary/go-to-market-gtm-strategy
11. Futter, D. (2020, May 27). Early beginnings – The rise of Xiameter. LinkedIn. www.linkedin.com/pulse/early-beginnings-rise-xiameter-daniel-futter
12. Singh, S., Maheshwari, S., & Mantri, A. (2023, August 13). Leading the future of India's petrochemical industry: Five winning principles for 2047. Indian Chemical News. www.indianchemicalnews.com/opinion/leading-the-future-of-indias-petrochemical-industry-five-winning-principles-for-2047-18552
13. Economic Times. BASF solutions for improved safety in eMobility. (2021, December 07). Retrieved May 24, 2024 from www.plastech.biz/en/news/New-BASF-Polyurethane-applications-technology-center-near-St-6960
14. BASF. Development partnerships – Collaboration with customers and partners. (n.d.). Retrieved May 24, 2024, from www.basf.com/global/en/who-we-are/history/joint_success/development_partnerships.html
15. BASF. Performance polymers. (n.d.). Retrieved May 24, 2024, from https://plastics-rubber.basf.com/global/en/performance_polymers
16. BASF. Who we are. (n.d.). Retrieved May 24, 2024, from www.basf.com/in/en/who-we-are/strategy/our-brand.html#:~:text=the%20chemical%20industry.-,Our%20corporate%20purpose%20%E2%80%93%20We%20create%20chemistry%20for%20a%20sustainable%20future,core%20strengths%3A%20our%20Verbund%20concept.
17. BASF. Management consulting. (n.d.). Retrieved May 24, 2024, from www.basf.com/global/en/careers/professionals/inhouse-consulting/management-consulting/practices.html
18. BASF. Technical support. Retrieved May 24, 2024 from https://chemicals.basf.com/global/en/Petrochemicals/Plasticizers/technical-support.html
19. BASF. Culture of collaboration: Where driving forward begins. (n.d.). Retrieved May 24, 2024, from https://automotive-transportation.basf.com/global/en/automotive/learn/stories/collaboration-helps-drive-forward-together

20. BASF. Regional results. (n.d.). Retrieved May 24, 2024 from https://report.basf.com/2022/en/managements-report/regional-results.html
21. BASF. Automotive lightweighting: 3 industry-leading innovations. (n.d.). Retrieved May 24, 2024, from www.basf.com/tw/en/media/BASF-Information/Innovation/Worlds-lightweight-future
22. BASF. Performance polymers. (n.d.). Retrieved May 24, 2024, from https://plastics-rubber.basf.com/global/en/performance_polymers.html
23. BASF. Driving innovation. (n.d.). Retrieved May 24, 2024, from www.basf-coatings.com/global/en/shaping-the-future/driving_innovation.html
24. BASF. Innovative solutions for future cars. (n.d.). Retrieved May 24, 2024, from www.basf-coatings.com/global/en/products-and-solutions/automotive_oem_coatings.html
25. Dow. Technical library. (n.d.). Retrieved May 24, 2024, from www.dow.com/en-us/technical-content.html
26. ORLEN. ORLEN Group launching new petrochemical product. (2022, October 3). Retrieved May 24, 2024, from www.orlen.pl/en/about-the-company/media/press-releases/2022/october-2022/orlen_group_launching_new_petrochemical_product
27. Roseneft. Angarsk Petrochemical Company launched new product – liquid oxygen. (2021, May 27). Retrieved May 24, 2024, from www.rosneft.com/press/news/item/206531/
28. Bombay Stock Exchange. (n.d.). *Disclosure under Regulation 30 of the SEBI (Listing Obligations and Disclosure Requirements) Regulations, 2015*. Retrieved June 24, 2024, from www.bseindia.com/xml-data/corpfiling/AttachHis/74747c33-cb00-4cea-b7e1-0f6cb785e47f.pdf
29. Economic Times. (2023, July 31). Reliance jumps 16 places, now at number 88 on Fortune Global 500 list. *Economic Times*. Retrieved June 24, 2024, from https://economictimes.indiatimes.com/news/company/corporate-trends/reliance-jumps-16-places-now-at-number-88-on-fortune-global-500-list/articleshow/102360683.cms?from=mdr
30. Reliance Industries Limited. (n.d.). Products & Brands. *Reliance Industries Limited*. Retrieved June 24, 2024, from www.ril.com/about/products-brands
31. Indian Textile Magazine. (n.d.). Reliance Industries: Global leadership in man-made fibres. Indian Textile Magazine. Retrieved from www.indiantextilemagazine.in/reliance-industries-global-leadership-in-man-made-fibres/
32. The Economic Times. (2016, March 23). RIL in branding alliance with Tamil Nadu firm for Recron fiber brand. *The Economic Times*. https://economictimes.indiatimes.com/industry/indl-goods/svs/paper-/-wood-/-glass/-plastic/-marbles/ril-in-branding-alliance-with-tamil-nadu-firm-for-recron-fibre-brand/articleshow/51498316.cms?from=mdr
33. Moneycontrol News. (2019, July 4). RIL partners with Turkey's Kivanç Tekstil to manufacture and market eco-friendly textile. Moneycontrol. www.moneycontrol.com/news/business/companies/ril-partners-with-turkeys-kivanc-tekstil-to-manufacture-and-market-eco-friendly-textile-4183261.html
34. Fibre2Fashion News Desk. (2019, May 9). RIL, Gimatex to co-brand Recron SuperBlack yarn. Fibre2Fashion. www.fibre2fashion.com/news/textile-news/ril-gimatex-to-co-brand-recron-superblack-yarn-249711-newsdetails.htm

35. Fibre2Fashion News Desk. (2005, May 9). RIL, Gimatex to co-brand Recron SuperBlack yarn. Fibre2Fashion. www.fibre2fashion.com/news/home-furnishings-news/newsdetails.aspx?news_id=2284
36. CNBC-TV18. (2023, June 20). RIL's Recron FS polyester gets sustainable, enhanced fire-retardant properties with Nofia technology. CNBC-TV18. www.cnbctv18.com/business/companies/ril-recron-fs-polyester-sustainable-enhanced-fire-retardant-properties-nofia-technology-15058181.htm
37. FRX Innovations. (2023, June 20). Reliance to enhance fire-retardant and sustainable properties of Recron FS using FRX Innovations' Nofia technology. FRX Innovations. www.frx-innovations.com/news/reliance-to-enhance-fire-retardant-and-sustainable-properties-of-recron%C2%AE-fs-using-frx-innovations%E2%80%99-nofia%C2%AE-technology
38. Fashion Network. (2017, February 13). Precot to use RIL's Recron SHT in sewing threads. Fashion Network. https://in.fashionnetwork.com/news/Precot-to-use-ril-s-recron-sht-in-sewing-threads,673175.html#chanel
39. CNBC-TV18. (2023, June 20). RIL's Recron FS polyester gets sustainable, enhanced fire-retardant properties with Nofia technology. CNBC-TV18. www.cnbctv18.com/business/companies/ril-recron-fs-polyester-sustainable-enhanced-fire-retardant-properties-nofia-technology-15058181.htm
40. Textile Value Chain. (2023, June 25). RIL's Recron SuperBlack fibers create unprecedented attraction at HEP meet in Bhilwara. Textile Value Chain. https://textilevaluechain.in/news-insights/industry-cluster-news/textile-corporate-sme-news/rils-recronr-superblack-fibers-create-unprecedented-attraction-at-hep-meet-in-bhilwara
41. Indian Textile Magazine. (n.d.). RIL partners Gimatex Industries to co-brand Recron SuperBlack yarn. Indian Textile Magazine. Retrieved from https://e-itm.net/wp-content/uploads/2019/05/RIL-partners-Gimatex-Industries-to-co-brand-Recron%C2%AE-SuperBlack-yarn.pdf
42. Fortune. (n.d.). *Indian Oil Corporation Limited*. Retrieved June 24, 2024, from https://fortune.com/company/indian-oil/
43. Indian Oil Corporation Limited. (n.d.). Propel brochure. www.iocl.com/download/Propel%20Brochure_new.pdf
44. Indian Oil Corporation Limited. (n.d.). IndianOil showcased its full slate of petrochemical product under the brand PROPEL [Video]. Facebook. www.facebook.com/IndianOilCorpLimited/videos/indianoil-showcased-its-full-slate-of-petrochemical-product-under-the-brand-prop/2030172143674411/
45. Rosalina, V., Natsir, M., & Suhendarsah, A. (2013). Customer relationship management (CRM) analysis at a petrochemical company. *International Journal of Engineering Associates*, 2(7). ISSN 2320-0804.
46. Zhejiang Petrochemical Co., Ltd. (2023). Establishment of CRM system in ZPC. www.zpcir.com/en/news/104/Establishment-of-CRM-System-in-ZPC
47. Sanayei, A. (2004). Electronic commerce customer relationship management: A case study of the petrochemical industry. *Iranian Journal of Information Science and Technology*, 2(2), 21–30.
48. Loop Horizon. (n.d.). CRM transformation for a global petrochemicals company. www.loophorizon.com/case-study/crm-transformation-for-a-global-petrochemicals-company

49. Udesk Global. (n.d.). Provides ExxonMobil with fast and convenient services center for customer. Udesk. Retrieved May 29, 2024, from www.udeskglobal.com/customer-case/exxon-mobil
50. Comarch. (n.d.). Devin Miller from ExxonMobil about loyalty strategy and customer experience trends [Podcast]. Comarch. Retrieved May 29, 2024, from www.comarch.com/podcasts/devin-miller-from-exxonmobil-about-loyalty-strategy-and-customer-experience-trends/
51. Touchpoints GmbH. (n.d.). BASF. Retrieved May 29, 2024, from www.touchpoints.ch/en/basf
52. Ha, T. (2014, April 6). CRM at Dow Chemical case study. Business Case Studies. Retrieved May 29, 2024, from https://business-cases-studies.blogspot.com/2014/04/crm-at-dow-chemical-case-study.html
53. Oracle. (n.d.). Dow: Customer success. Oracle. Retrieved May 29, 2024, from www.oracle.com/in/customers/dow-1-mrk-cl/
54. Salesforce. (n.d.). Shell Chemicals powers service revolution. Salesforce. Retrieved May 29, 2024, from www.salesforce.com/eu/customer-success-stories/shell-chemicals/
55. Koskinen, P., & Forbes Technology Council. (2021, April 29). Personalization: How to make a B2B brand stand out and thrive. Forbes. www.forbes.com/sites/forbestechcouncil/2021/04/29/personalization-how-to-make-a-b2b-brand-stand-out-and-thrive/?sh=1d4d4aae2866
56. Coveo. (2023, February, 14). Dow's digital acceleration: Creating an end-to-end experience. Retrieved May 24, 2024, from www.coveo.com/blog/dows-digital-acceleration-strategy/
57. Materialsperformance.com. (2020, October 9). Dow polyurethanes accelerates its digital transformation. Retrieved May 24, 2024, from https://materialsperformance.com/news/2020/09/dow-polyurethanes-accelerates-its-digital-transformation
58. Dow. Harnessing the power of digitalization. Retrieved May 24, 2024 from www.dow.com/en-us/category/product-technology/pt-polyurethanes/harnessing-the-power-of-digitalization.html

5 HSE Management and ESG Compliance in Petrochemicals Business

5.1 INTRODUCTION

As discussed in previous chapters, petrochemical business is a high-end technology intensive business that involves multiple processes in a multifaceted context. In addition to that, it is also a high energy-intensive business.

Given the complexities of the business that involve an interface of man, materials, energy, and technology, adequately managing health, safety, and environment (HSE) issues are vital at every step of the value-chain of the petrochemical industry.

This chapter thus deals with HSE management and management of environmental, societ, and governance (ESG) issues of the petrochemical business. In the upcoming section at the beginning, we provide an overview of key safety hazards in the refining and petrochemical industries.

5.2 SAFETY HAZARDS RELATED TO PETROCHEMICAL INDUSTRY

Petrochemical industry is a complex process industry and thus due to the sheer nature of the (toxic, inflammable, etc.) materials handled and technological processes involved are characteristically hazardous also. We shall discuss below some major safety hazards associated with the industry:

1. **Fire and Explosions:** The processes of the industry involve handling of inflammable materials like crude oil, gasoline, and various other chemicals. This increases the risk of fires and explosions, which can, for example, be triggered by sparks (welding, electric circuit, etc.), malfunction of equipment, and even human errors also.
2. **Toxic Substances' Releases:** Petrochemical units handle a wide range of hazardous chemicals like benzene, hydrogen sulfide, sulfur dioxide, and various other volatile organic compounds (fumes). Improper handling or treatment can result in serious health risks if they get released into the environment and/or if workers are exposed to them.

DOI: 10.1201/9781032661933-5

3. **Process Safety-Related Incidents:** Non-adherence to safe processes, etc. negligence at the petrochemical plants may result in disastrous accidents, such as fires, explosions, and toxic releases.
   See Figure 5.1.
4. **Pressure-Related Accidents:** Petrochemical manufacturing processes operate under high temperature and pressures. This requires the use of equipment like pressure vessels, which malfunction due to various reasons, including lack of maintenance, corrosion, hyper-pressures, or damage to metal-integrity, etc. This in turn may lead to leaks, ruptures, or explosions type of accidents.
5. **Exposure to Chemicals:** The petrochemical industry workplace conditions have the potential of hazardous chemical exposure to workers that may occur by ingestion, skin contact, or inhalation. This can have serious health repercussions, such as cancer, skin diseases, and/or respiratory diseases.

FIGURE 5.1 Petrochemical Industry Hazards.

6. **Hazards Related to Confined Spaces:** Petrochemical unit processes involve numerous tasks that involve working in confined spaces such as storage tanks, reactors, tankers, containers, etc. Working conditions in such closed/covered spaces can are exposed to the risks of entrapment, asphyxiation, and /or toxic gas inhalation, etc.
7. **Temperature Hazards:** Workers of petrochemical units work under very high temperature conditions, which could result in heat-related illnesses such as heat fatigue or heatstroke, etc.
8. **Manual Operations and Ergonomic Hazards:** Workers of petrochemical plants are very often required to lift heavy equipment. In addition, they also carry out many repetitive tasks, and are subjected to difficult postures during work. Thus, if proper ergonomic principles applicable to manufacturing workplaces are not applied then it can lead to muscle/orthopedic-related injuries among them.
9. **Noise and Vibration:** The petrochemical workplace environment is subjected to high decibels of sound as well as vibration coming from machines like pumps and compressors. If adequate precautions are not taken, then this may cause medical conditions like hearing loss and other disorders among the workers.
10. **Natural Disasters and External Events:** Like any other manufacturing/process industry, petrochemical units are also exposed to natural calamities such as typhoons, earthquakes, tsunamis, floods, etc. At the same time, they may also face terrorist attacks or sabotage. All of these have the potential to cause extensive damage and are thus serious safety hazards.

Petrochemical industry needs to realize all the types and extent of all such hazards and need to have a structured HSE framework in place so as to mitigate these hazards. The HSE framework, apart from a strong safety culture, should also include stringent safety regulations, wide-ranging training programs, robust engineering and instrumentation controls, and effective emergency response plans among others.

After having discussed the hazards that the petrochemical industry is exposed to, it is contextual to discuss the HSE issues of the petrochemical industry in the below section.

5.3 HSE ISSUES RELEVANT TO THE PETROCHEMICAL INDUSTRY

Given its nature of operations, the petrochemical industry faces various HSE issues, and the same have significant implications for all stakeholders especially the workers, affected communities, and environment. We now discuss some of the critical HSE issues in the petrochemical sector as follows:

1. **Workplace Health and Safety:** As also briefly discussed above, the petrochemical units may present risks to worker health and safety due resulted by exposure to hazardous chemicals and toxic substances/fumes, risks of fires, and explosions. Workplace safety issues also include possible accidents during manufacturing, logistics, and maintenance.
2. **Chemical Exposure and Health Risks:** Workers may be exposed to harmful substances through inhalation, skin contact, or ingestion, leading to respiratory issues, dermatitis, neurological disorders, and long-term health complications such as cancer and reproductive effects.
3. **Environmental Pollution:** Emissions, wastewater discharges, and waste-disposal practices from petrochemical facilities can result in environmental pollution, ecosystem contamination, and adverse effects on air and water quality, biodiversity, and ecosystems due to the release of pollutants into the air, water, and soil.
4. **Process Safety Management:** Petrochemical processes involve intricate operations, high temperatures, pressures, and reactive chemicals, necessitating robust systems for managing process safety to prevent accidents, releases, and calamitous incidents like chemical spills, oil spills, fires, and explosions.
5. **Climate Change:** The activities in the petrochemical industry result in the generation of greenhouse gas emissions due to energy consumption, flaring, and venting of gases. Additionally, the production of carbon-intensive products like plastics further contributes to these emissions, impacting extreme weather, rising sea levels, and ecological damage while also playing a role in causing climate change and global warming.
6. **Emergency Readiness and Response:** Petrochemical companies must have robust emergency and response plans to mitigate the impacts of accidents, spills, and emergencies, protect workers, communities, and ecosystems, and ensure effective communication, coordination, and cooperation with stakeholders (particularly in the vicinity) and emergency responders.
7. **Health and Safety of the Local Community:** Nearby petrochemical facilities can bring about potential risks to the health and safety of local residents through exposure to pollutants, accidents, and emergencies, as well as causing worrics about noise, odors, and visual disturbances.
8. **Waste Management:** Within petrochemical operations, various wastes such as wastewater, solid waste, and hazardous waste are generated, necessitating proper management, treatment, and disposal to prevent contamination, pollution, and detrimental effects on the environment and public health.
   See Figure 5.2.

Addressing these HSE issues requires a comprehensive approach that involves regulatory compliance, risk assessment, hazard identification, engineering controls, training and education, monitoring and surveillance, emergency planning, stakeholder engagement, and continuous improvement efforts to enhance safety,

FIGURE 5.2 Petrochemical Industry HSE.

protect human health, and minimize environmental impacts in the petrochemical sector. Before discussing how the petrochemical industry tackles the HSE challenges, it is contextual to know about some historical disasters due to non-observance of adequate HSE management practices in the past.

5.4 SOME MAJOR ACCIDENTS IN THE GLOBAL PETROCHEMICAL/CHEMICAL/OIL AND GAS INDUSTRY HISTORY AND THE HSE PERSPECTIVE[1]

Unfortunately, there have been some major accidents in the history of the global petrochemical/chemical/oil and gas industries and experts have put those in the perspective of HSE management practices. Here are some notable examples:

1. **Bhopal Disaster (1984)**: While not directly in the petrochemical industry, the Bhopal disaster involved a pesticide plant owned by Union Carbide Corporation (now part of Dow Chemical Company) in Bhopal, India. The release of methyl isocyanate (MIC) gas resulted in thousands of immediate fatalities and long-term health effects for tens of thousands more. The disaster was primarily attributed to inadequate safety systems, poor maintenance, and lack of emergency preparedness [1].
2. **Piper Alpha Disaster (1988)**: 167 people died in a disastrous explosion and fire that occurred on the Occidental Petroleum-operated Piper Alpha site in the North Sea. A gas leak caused the incident; it ignited and resulted in several explosions. Failures in safety systems, inadequate risk

assessment, and poor communication among personnel were identified as contributing factors [2].

3. **Texas City Refinery Explosion (2005)**: At the BP Texas City Refinery, there was an explosion and fire that left over 170 people injured and 15 dead. The occurrence was brought on by a raffinate splitter tower overflowing, releasing combustible hydrocarbons that caught fire. The US Chemical Safety and Hazard Investigation Board (CSB) determined that organizational culture, mechanical integrity, and process safety management shortcomings were the major issues [3].
4. **Deepwater Horizon Oil Spill (2010)**: The Deepwater Horizon explosion and accompanying oil spill in the Gulf of Mexico, although predominantly an offshore drilling accident, had a major impact on the petrochemical industry. BP and its contractors were held accountable for a sequence of mishaps including well management, safety systems, and emergency response readiness, culminating in the greatest marine oil leak in history and 11 fatalities [4].
5. **Tianjin Explosions (2015)**: In Tianjin, China, a series of explosions occurred at a warehouse storing hazardous chemicals, including sodium cyanide and ammonium nitrate. The blasts resulted in at least 173 fatalities, hundreds of injuries, and widespread damage to surrounding areas. These explosions heightened concerns about the aspects of safe storage and handling of hazardous chemicals in a petrochemical plant. These also underlined the wide gaps in implementation of regulations and requisite emergency response capabilities of the petrochemical companies [5].

   All the critical incidents discussed above underline the significance of putting the aspects of safety at top priority. They also bring to the fore the need for adequate implementation of desirable HSE management systems, along with adherence to best (HSE management) practices. Only such an implementation can prevent calamitous accidents and protect the health and welfare of the workers, local communities, and the ecological environment. Discussion and analysis of such critical incidents is important to increase the awareness about augmenting scrutiny, introduce regulatory reforms, and also petrochemical industry-wide measures to improve safety performance so that similar tragedies' recurrence can be prevented in the future.

The upcoming section shall discuss the ways in which the petrochemical industry addresses the HSE challenges.

5.5 TACKLING OF HSE CHALLENGES BY THE PETROCHEMICAL INDUSTRY

The petrochemicals industry, globally, employs various strategies and initiatives to tackle the HSE challenges. We discuss below the broad ways in which the industry attempts to overcome these challenges:

1. **Compliance with HSE Regulations**: Petrochemical companies take all measures (including those related to human resources, investments, and technology) to adhere to all relevant (local, national, and international) HSE regulations and standards. By doing this the petrochemical companies make sure that their manufacturing operations comply with basic minimum requirements for HSE management practices.
2. **Risk Evaluation and Management**: It is normal (and desirable also) for petrochemical companies to undertake comprehensive risk assessments so as to identify all possible hazards and risks associated with their plant operations. A structured risk evaluation process enables them to implement risk management strategies that lead to mitigation of prospective hazards, taking prevent measures to avoid accidents, and also lessen the probability and impact of any such (unfortunate) incidents.
3. **Process Safety Management**: Once risks are evaluated, it helps the petrochemical plants to implement adequate process safety management systems aimed at avoiding (any) major accidents, such as fires, explosions, and chemical/toxic releases/fumes. Process safety management, among others, also deals with hazard identification, risk analysis, and (rapid) emergency shutdown procedures.
4. **Training and Education**: While technology and expenditures have their own importance, petrochemical companies are constantly mindful of creating a structured HSE learning system for their employees. Human Resources function of petrochemical companies design programs that inculcate awareness and advanced learning of HSE risks and promoting safe workplace practices. Very often all training towards these cover aspects like handling of hazardous fluids, quick emergency response, proper usage of personal protective equipment (PPE), and taking the lead in environmental initiatives.
5. **Investment in Technology and Innovation**: The petrochemical industry realizes the imperative of investing in technology and innovation so that safer and more environmentally friendly processes, products/services, and technologies are designed and implemented. The measures towards this comprise sophisticated process automation, (computer/AI based) control systems, (remote/in-situ) monitoring equipment, and global/region-specific technologies for emission control and effluent treatment.
6. **Environmental Management Systems (EMS)**: The ISO has developed relevant codes for management of the environment. Globally, all petrochemical companies implement, in a structured manner, these ISO-EMS to manage and enhance their environmental performance. EMS frameworks such as ISO 14001 deal with identification of desirable/essential environmental norms, setting of objectives, monitoring the implementation, and review of progress along with feedback loops at all stages. All petrochemical companies have not only adopted ISO 14000 but they also implement it with commitment.

7. **Pollution Control and Waste Management**: Globally, the petrochemical industry implement measures to reduce pollution and minimize waste (of all types) over the entire manufacturing life-cycle. Very often this includes process-optimization, enhancing efficiency, reduce, recycle, and reuse (3R) initiatives, and implementing waste/effluent treatment (and also safe disposal) technologies.
8. **Collaborative Approach**: Worldwide the petrochemical companies realize the need to engage with (potentially affected) local communities, government organizations, NGOs, and all other relevant stakeholders in order to alleviate HSE concerns, fostering trust and confidence, and promoting transparency and integrity. Towards this objective, local community outreach programs are conducted. In addition, stakeholder consultations/public-meetings are also organized to collaborate everyone on HSE issues.
9. **Continuous Improvement (Kaizen)**: Quality management practices such as *Kaizen* are undertaken by the petrochemical companies so that continuous improvement towards HSE management is ensured and also the companies can track their own HSE performance. This helps in identifying areas for improvement, if any, and takes corrective actions.

As seen, the global petrochemicals industry with the implementation of these strategies and measures, work towards mitigating HSE risks, safeguarding the workplace, local communities, and ecosystems. These strategies also aid in promoting sustainable operations and inculcate a sense of ownership of utilizing natural resources. As discussed, the petrochemical industry globally adopts a range of structured approaches to identify possible risks, and operational issues in their operations along with mitigating/tackling the same. The upcoming section discusses one of these important methods.

5.6 HAZOP IN THE PETROCHEMICAL SECTOR

HAZOP, the abbreviation for "Hazard and Operability Study", is a structured approach worldwide to identifying all possible risks and operability (functioning with safety and reliability) issues in industrial processes, including those contextual to the petrochemical sector as well. In the discussion below, we take a bird's eye view of HAZOP in the context of the petrochemical companies:

1. **Objective of HAZOP**: A HAZOP study aims at systematically examining a process or system to identify possible deviations from its (originally) intended design and/or operations that could result in hazards, accidents, bottlenecks, or operability glitches.
2. **Approach**: HAZOP is usually undertaken by a multifunctional team of experts, which may consist of process engineers, shop floor operators, safety experts, and other pertinent stakeholders. This team works towards

identifying (any) potential deviations (termed as *nodes*) from (predefined) standard operating procedures (SOPs) by examining every component of the process, including operations, process control systems, instrumentation, and equipment (handling/maintenance, etc.) in a systematic manner.

3. **Identification of Problem-Areas**: In this step, the concerned team identifies the problem by systematically iterating process parameters such as flow rates, temperatures, pressures, compositions, etc. For each potential problem, the team examines probable deviations in terms of parameters such as overpressure, over-temperature, underflow, overflow, loss of containment, and other abnormal conditions. These deviations lead to the problem-areas, which are then rectified.
4. **Scenario Planning**: Subsequent to the identification of problem-areas, the team creates certain scenarios in order to evaluate the consequences of all such problems in terms of safety, product-quality, environmental impact, operations, etc. Some examples of potential problem-areas are the release of hazardous materials, equipment malfunction, production-losses, impact on workers' safety, etc.
5. **Risk Assessment**: With an aim to assign ranks to particular risks and therefore decide suitable risk mitigation measures, the expert team evaluates the risk associated with each identified problem-area based on various gauges such as their probability of occurrence, severity of impact, etc.
6. **Expert Recommendations**: In the next step, based on the results of the HAZOP investigation, the expert team provides recommendations to deal with the identified problem-areas and also extends suitable suggestions. The suggestions could be for modifying designs, procedure-rationalization, and revamping the safety systems, among others.
7. **HAZOP Process Documentation**: Finally, a HAZOP report containing the HAZOP study's findings, recommendations, and actions taken to mitigate the identified risks arising out of problem-areas is prepared. The report is an important reference document for process safety management, risk assessment/mitigation, and decision-making, and is also for the purpose of all future references.

Therefore, in a nutshell, it could be inferred that the petrochemical industry in order to proactively identify and mitigate hazards, augment process safety, and improve the industrial processes, uses HAZOP as a critical tool. HAZOP with the aim of minimizing the risk of accidents and critical incidents aids the petrochemical companies to ensure the safe and consistent operations of their plants [6].

As deduced from the above discussions, prevalence of a good safety culture is crucial to adequate management of HSE in the petrochemical industry. We thus discuss below the (soft) issue of safety culture in petrochemical companies. See Figure 5.3.

FIGURE 5.3 HSE Risk Management.

5.7 ROLE OF ORGANIZATIONAL CULTURE IN MANAGING HSE IN THE PETROCHEMICAL INDUSTRY

As discussed above, we now discuss the salient aspects about the role of organizational culture in managing HSE in the petrochemical industry. Prominent amongst these aspects are values, leadership, empowerment, communications, learning, incentives, and integrative approach. We take them up one by one.

1. **Inculcation of Safety as a Core Value**: Any organization's commitment to safety as a core value is crucial to building a strong HSE culture. It is natural that wherever safety is accorded a top priority and deemed non-negotiable, it is percolated to all levels of an organization and its activities, and also in all operational and strategic processes.
2. **Leadership Commitment**: It's the leadership's *buy-in* that plays an important role in establishing a safety culture in an organization. Senior management's strong belief ensures alignment of all other employees towards HSE, and they become increasingly inclined to adopt safety in their work with serious commitment. Exemplary leadership also helps in providing adequate resources for all HSE activities.
3. **Employee Engagement and Empowerment**: Any organization, in order to see HSE culture to be effective, ensures the active engagement of employees at all organizational levels. Employees are encouraged to take initiatives and mistakes if any are also taken in right earnest by the management. They are also empowered to not only extend suggestions about hazards, but also suggest/execute measures that improve safety. Employee engagement and empowerment ensures a deep sense of ownership amongst them towards HSE.

4. **Openness in Communication and Transparency**: Openness in communication and transparency regarding HSE are important in fostering a culture of safety in organizations. These aspects should result in an organizational culture and atmosphere where the employees should be candid enough in conveying their safety concerns without any fear of reprisal. Openness in communication also means that the management extends relevant feedback (in time) and apprises the employees on safety initiatives and performance. Trust building and collaboration on HSE issues increases with the prevalence of transparent communication.
5. **Continuous Learning and Improvement**: A good safety culture in organization considers any investment in learning and development towards HSE aspects as an investment and not an expenditure. That's because organizational decision makers are aware that adequate (and contemporary) skills and knowledge are needed by the employees to dispense with their jobs safely.
6. **Incentives and Recognitions**: A good organizational culture believes in incentivizing its people for their contributions to the organization's HSE performance. Recognition and incentives, cement positive safety behaviors among the employees and inspire their consistent commitment to safety.
7. **Integration of HSE Into Organizational Processes**: A healthy organizational HSE culture over a period of time by default results into, integration of key business activities including planning, design, procurement, logistics, R&D, operations, etc. towards the concerns of safety, environment, and well-being. This also results in the organizations getting HSE risk signals early on so that these risks can be tackled proactively, leading to improvement of overall safety performance.

As discussed above, organizational culture plays an important role in managing HSE by the petrochemical companies as a good culture molds attitudes, behaviors, and practices related to safety amongst its people. A strong HSE culture based on positive values, leadership's belief, employee empowerment, transparent communications, fostering learning, providing incentives and considering safety as an integrative approach results in an organization that is at apex of its HSE practices.

5.8 ISO CODES RELEVANT TO HSE PRACTICES IN THE PETROCHEMICALS INDUSTRY

Similar to many other ISO standards for other functions, HSE management in the petrochemical industry also aligns with some prominent ISO standards like those pertaining to occupational health, environment, quality, energy efficiency, and information systems' security. These ISO codes relevant to HSE in the petrochemical industry are discussed in the below section as follows:

1. **ISO 45001: Occupational Health and Safety Management Systems**: This code pertains to designing, implementing, maintaining, and continuously improving an occupational health and safety management system (OHSMS) in organizations, including petrochemical companies. Apart from supporting organizations in ascertaining and managing risks to people's health and safety, ISO 45001 also helps the companies towards maintaining legal and regulatory compliance [7].
2. **ISO 14001: Environmental Management Systems**: As also discussed in a previous section, ISO 14001 helps organizations delineate the requirements for an effective EMS that enables them effectively to manage the environmental impact of petrochemical companies' operations. Further, this standard also helps the organizations in the design and implementation of strategies to alleviate environmental impacts, compliance with environmental laws, and consistent delivery of environmental performance [7].
3. **ISO 9001: Quality Management Systems (QMS)**: While quality management is at the core of ISO 900, it also contains sub-codes which are contextual to HSE management in organizations. Further organizations, by implementing an ISO 9001-certified QMS, can demonstrate and communicate about their processes' alignment with quality, safety, and environmental objectives, aimed at augmenting overall performance [7].
4. **ISO 50001: Energy Management Systems**: Petrochemical companies, by their characteristics, are energy intensive in their operations. And hence they have a scope for bringing about energy efficiency in all of their processes. ISO 50001 offers a framework for designing, implementing, maintaining, and improving EnMS that facilitates petrochemical companies to continuously explore opportunities for energy savings, energy transition, reduce energy costs, and mitigate the negative environmental effects of energy consumption [8].
5. **ISO 27001: Information Security Management Systems (ISMS)**: HSE-related information is sensitive in nature and may lead to complications and panic, if not handled adequately. Petrochemical companies, due to their handling of a variety of environmentally sensitive materials are particularly vulnerable to data breach events. ISO 27001, that pertains to designing, implementing, maintaining, and continuously improving an ISMS ensures the data privacy, data integrity, and also integrity of a petrochemical company's information assets, documentation, and record keeping [9, 10].

Petrochemical firms can boost their HSE performance, reduce risks, ensure regulatory compliance, and exhibit their commitment to responsible business practices by implementing and upholding compliance with these ISO standards.

Having understood the HSE management aspects of the petrochemical industry, it would be contextual to see how these practices are applied in a company specific context. The following section discusses HSE management at an Indian petrochemical company.

5.9 HSE MANAGEMENT: AN ILLUSTRATION OF OPAL (ONGC PETRO ADDITIONS LIMITED)

Company Overview: An Indian petrochemical company, ONGC Petro additions Limited (OPaL; www.opaliindia.in) was established as a joint venture of Indian public sector oil and gas company ONGC Limited (www.ongcindia.com) and other Indian energy companies. In 2023, it had a turnover of USD 43.6 million [11]. As of 2024, their petrochemical complex spans 5 square kilometers and has the capacity to produce 1.4 million tons of polymers and 0.5 million tons of chemicals, including 400 KTPA of propylene and 1,100 KTPA of ethylene. Related units include a butadiene extraction unit, benzene extraction unit, and pyrolysis gasoline hydrogenation unit. There are two 360 KTPA LLDPE/HDPE swing units, one 340 KTPA dedicated HDPE unit, and one 340 KTPA PP unit in the polymer factories [11].

The petrochemical complex enjoys a competitive edge due to its strategic location in the port city of Dahej, Gujarat-India, which provides it with access to both domestic and international markets [12].

OPaL has structured processes in place to implement best practices in HSE. It employs an "Integrated Management System Policy" (Quality, Environment, Health, Safety, and Security) to ensure this.

As per their public declaration, by institutionalizing a strong Integrated Management System and maintaining a constant focus on meeting the expectations of their customers and stakeholders while ensuring the best possible use of natural and other resources, ONGC Petro additions Limited (OPal) produces and supplies petrochemical products and related services to both domestic and international markets. Along with their organizational objective and vision, OPal believes that the preservation of the environment and ecology, as well as the safety and occupational health of their stakeholders, will always be one of their core principles.

To achieve this, OPaL aims at [13]:

- Focusing on the requirements and expectations of stakeholders and customers.
- Adhering to all relevant laws and any particular pledges it may make.
- Identifying short-and long-term goals, implementation plans, procedures, and adherence to global industry practices and standards in order to anticipate hazards and take appropriate action.
- Encouraging process-based thinking and risk-based thinking in all areas to boost individual accountability, performance, and ownership.
- Integrating the needs for energy management, environmental protection, and sustainable development; removing or preventing hazards to one's health at work; safety and security threats to individuals, property, and the

environment; safe working conditions; and safe behavior in all whatever the company does.

- Ensuring active consultation and engagement from workers, employees, and stakeholders.
- Giving workers and employees the knowledge, tools, and training necessary for environmental management, high-quality operations, and safe and secure working conditions.
- Maintaining awareness and preparedness for any emergency through the use of suitable action plans.
- Setting up efficient internal and external audit procedures to guarantee performance assurance and continuous improvement.

From the above it could be derived that OPaL (ONGC Petro additions Limited) aligns with the due processes like creating a safety culture, having proper HSE management systems in place, emergency response systems, and stakeholder engagement to ensure the implementation of its HSE policy [14].

At the same time, it is implicit that, as per the best practices prevailing in the petrochemical industry, OPaL also in its overall HSE management, must be observing the desirable techniques of Risk Assessment and Hazard Identification, and Incident Management [15].

By implementing these measures and initiatives, OPaL demonstrates its commitment to managing HSE risks effectively, protecting the health and safety of its employees and contractors, and minimizing its environmental footprint while delivering value to its stakeholders. Towards that, it has also received accolades from prestigious agencies for its commitment to environment management [16].

The discussion on HSE management shall be concluded with some more company-specific examples from a global perspective.

5.10 BEST HSE MANAGEMENT PRACTICES IN THE GLOBAL PETROCHEMICALS INDUSTRY: SOME ILLUSTRATIONS

We give below an account of a few prominent petrochemical companies known for their exemplary HSE management practices that are worth emulating not only by other petrochemical companies but also these can be used as benchmarked by other manufacturing and/or process organizations. These are as below:

1. **Saudi Aramco**: Saudi Arabia headquartered Saudi Aramco is one of the world's prominent, integrated (present in all stages of value-chain) energy and petrochemical companies, and is also distinguished for its exemplary HSE management practices. The company is known for its robust safety management systems. The company has stringent safety procedures in place, conducts regular safety audits, and offers comprehensive HSE learning and development programs for its employees. Saudi Aramco's leadership is known for committing adequate investments in

technology and innovation to augment process safety and environmental performance [17].

2. **Shell**: London headquartered multinational energy and petrochemical company, Shell is recognized globally for its safety and environmental initiatives in the petrochemical industry. The company has famously launched the unique the *"Goal Zero"* program, which is a proactive approach to HSE management, focusing on zero harm to people and ensuring zero leaks [18]. With initiatives like these, Shell exemplifies stewardship, employee engagement, and continuous improvement towards HSE performance by also nurturing a safety culture across organization.
3. **ExxonMobil**: ExxonMobil, headquartered in Texas, USA, is also known for its robust process safety and risk management practices in the petrochemical sector. Aimed at preventing major hazard incidents and protecting the health and safety of its employees and also looking at safeguarding the environment, the company has designed and implemented comprehensive process safety management systems. These systems include risk assessment methodologies, safety integrity programs, and also rapid emergency response mechanisms. ExxonMobil believes in a collaborative culture and thus often teams up with its industry partners and stakeholders. This collaboration involves sharing best HSE practices, learning from others towards the objective of attaining the best HSE performance [19].
4. **Dow**: Dow Chemicals of the USA is known for its sustainability and environmental best practices in the petrochemical industry. The company follows a structured approach towards these practices and thus has also implemented necessary EMS, including the ISO 14001 certification [20]. The certification aids in minimizing its (negative) environmental impacts and at the same time, promotes resource conservation efforts. By continuously investing in innovative technologies and practices to reduce carbon and other emissions, energy management, and reducing waste across its global operations, Dow attempts to remain at the forefront of its HSE management practices [21].
5. **TotalEnergies**: TotalEnergies headquartered in France is recognized for its rounded approach to HSE management in the energy and petrochemical industries. The company adequately integrates HSE protection and concerns for community into its business operations. TotalEnergies has set ambitious HSE performance targets that include reducing greenhouse gas emissions and also enhancing safety performance. The company documents all the good practices and reports on its progress to its stakeholders on a regular basis. Total also has a practice of engagement with local communities and stakeholders to deal with HSE concerns and thus promote sustainable development in the all regions where it operates [22].

As observed, the above companies exhibit exemplary HSE management practices by demonstrating their commitment to safety, and dispensing with their environmental and social responsibilities in the petrochemical sector. The implementation of best practices and continuous improvement in HSE management, these companies contribute to safer workplaces, environmental protection, and also sustainable manufacturing practices in the global petrochemical industry.

After a comprehensive discussion on HSE management, it would be contextual to discuss the important aspect of ESG norms for the petrochemical industry in the below sections.

5.11 ENVIRONMENTAL, SOCIAL, AND GOVERNANCE (ESG) PARAMETERS FOR PETROCHEMICAL COMPANIES

Environmental, social, and governance (ESG), is a set of parameters, viz. environmental issues, social issues, and corporate governance in business organizations that can be considered in overall business performance and also for investing decisions in organizations on the basis of their ESG performance. Investing on the basis of ESG parameters is a desirable contemporary practice which is sometimes referred to as responsible investing or, impact investing. We briefly discuss these parameters as follows:

1. **Environmental**: As also referred to earlier, the environmental parameter assesses the impact of a business's operations on the ecological environment. Aspects like resource efficiency, energy management, waste reduction, and tackling carbon emissions constitute this parameter. Given the nature of their operations, the environmental concerns for petrochemical firms typically hinge on activities like greenhouse gas emissions, air, and water pollution, and the sustainability of their manufacturing/process operations.
2. **Social**: It is important to assess any business's overall interaction and dealings with the societal framework (employees, local communities, and other society stakeholders) that it operates in. The social dimension includes aspects like community engagement, workplace safety, multiculturalism, and diversity, labor practices, and human rights. Apart from general aspects like worker safety and community health, depending upon regional dispersion, the social dimension in the petrochemical sector may also include concerns of marginalized or indigenous groups.
3. **Governance**: Corporate governance refers to how a business organization is led and managed. Governance practices also help understand better how leadership's interests are aligned with stakeholder expectations, whether shareholder rights are protected, and the types of internal controls to uphold transparency and accountability on the part of corporate leadership. Governance concerns in the petrochemical industry may also include handling conflicts of interest, adherence to rules and regulations, and ethical conduct of businesses.

   See Figure 5.4.

FIGURE 5.4 Environment, Social and Governance.

After understanding the concept of the ESG framework, we now discuss the significance of ESG framework for the petrochemical sector as follows:

1. **Risk Management**: The tackling of ESG issues is important for petrochemical companies towards identifying and mitigation of certain specific risks to their business, including any type of statutory penalties, damage to corporate reputation, and potential operational disruptions.
2. **Resource Efficiency**: Petrochemical companies, by using resources more wisely, producing less waste, and adopting energy management practices can thus be sustainable and also reduce costs.
3. **Investment, Market Access, and Reputation**: As discussed above, investors increasingly rely on good ESG practices to deploy their funds. Similarly, customers and other stakeholders also often put the companies under ESG performance lens. Thus, exhibiting a strong ESG performance has the potential to enhance a petrochemical company's access to capital, market opportunities, and corporate-partnerships. All these also have reputational benefits for the company.
4. **Competitive Edge through Sustainability**: In a dynamic global economy, it is important for the petrochemical companies to adopt best practices so as to be high on ESG scores. Sustainability is crucial to gain a long-term competitive edge by the petrochemical companies.

As seen above, ESG norms' compliance is critical for petrochemical companies so that they are not only able to tide over the business and operational risks, but also enhances their overall reputation and competitive advantage. The global petrochemical industry endeavors to comply with ESG norms through various initiatives and practices, which are largely aimed at promoting sustainability, ethical business conduct, and engagement with stakeholders. The petrochemical industry largely addresses these ESG considerationsandby means of exhibiting a good environmental performance; incorporating innovative and sustainable technologies and materials in product development; complying with ESG regulations; deep engagement with various stakeholders; ensuring safety and health for workers and community; following ethics and transparency in business conduct and being accountable through a robust corporate governance mechanism. In general, it has been observed that the petrochemical industry indeed realizes the significance of the ESG framework towards their quest to creating long-term value, mitigating risks, and gaining buy-in of the stakeholders.

In order to have an applied perspective about how petrochemical companies integrate ESG principles into their business strategies and operations, we discuss a case of a prominent petrochemical company in the below section.

5.11.1 ESG: An Illustration of SABIC

SABIC {Saudi Basic Industries Corporation (www.sabic.com/en), a multinational petrochemical firm, has implemented the components of a significant ESG approach from the outset. As per SABIC's public declaration [23] the organization upholds careful and visionary corporate governance, a commitment to environmental stewardship, and a feeling of duty and concern for their larger stakeholders and society. SABIC believes in "setting an ambitious agenda for business growth towards becoming the preferred world leader in chemicals, these elements collectively represent the complex vision of sustainability that drives the company."

We discuss below how the company has demonstrated compliance with ESG norms:

1. **Environmental Performance**
   a. SABIC has made significant investments in technologies aimed at reducing its environmental impact. For example, it has developed innovative processes to increase energy and resources' efficiency [24] and reduce greenhouse gas emissions in its production facilities.
   b. SABIC has put in place wide-ranging recycling and waste management plans aimed at reducing waste-generation and promoting the circular economy. In order to introduce sustainability in fuel usage, SABIC uses renewable feedstocks and bio-based materials to curtail on (relatively polluting) fossil fuels [25].
2. **Social Concerns**
   a. SABIC lays a strong accent on employee health and safety by executing desirable safety standards, organizing learning and development programs, along with applying risk management systems across its business operations. Safety audits and inspections to identify and address potential hazards are undertaken by the company on a regular basis and are deeply ingrained in the organizational culture [26].

   b. SABIC's community development initiatives (CDIs) of investing in education, healthcare, safety-infrastructure, and community welfare aimed at local capacity development and good quality of life are unique in nature [27.]
3. **Ethical Governance**
   a. SABIC endeavors to implement and sustain high standards of corporate governance, ethics, and transparency. This is accomplished through an independent board of directors and a well-defined accountability mechanism. In order to follow ethical business practices the company has zero tolerance for corruption. At the same time, SABIC also complies with all relevant international conventions and laws for integrity and fairness in business [28].
   b. SABIC through its annual sustainability reports and a well-designed corporate communications framework, follows all information disclosure norms towards ESG performance. Those include relevant environmental metrics, social impact assessment reports (SIAs), and corporate governance practices. This level of transparency augments stakeholder-trust and overall company perception [29].
4. **Collaborative Culture**
   a. SABIC fosters a collaborative culture aimed at innovation and sharing the best ESG practices in the petrochemical sector. Towards this, it engages with its industry partners R&D institutions, and government agencies. The company is known for keenly engaging in relevant industry activities to address global issues of sustainable natural resources' usage, and net zero, including the use of sustainable materials and concerned technologies [30].

As seen, SABIC, by pledging to ESG principles and aligning with sustainable business practices, has carved a place in the petrochemical industry as a harbinger of sustainability, social champion, and ethical business.

Towards the end of the discussion on ESG, the below section discusses how some prominent petrochemical companies' exemplary ESG practices have been acknowledged and awarded accolades.

5.12 BEST ESG PRACTICES ACCOLADES FOR PETROCHEMICAL COMPANIES: SOME CASES

Having understood the significance and benefit of ESG for the petrochemical industry, it would be interesting to see how some global petrochemical companies have been recognized for their exemplary ESG practices through many awards and accolades, as follows:

1. **Saudi Aramco**: Owing to its commitment to sustainability and responsible business practices, Saudi Aramco has been recognized at a global level. For

instance, SABIC has received awards such as the prestigious S&P Global Platts Global Energy Awards for exemplary Corporate Social Responsibility (CSR), as an acknowledgment of its efforts to address environmental challenges, foster community development, and following best corporate governance practices [31].

2. **Dow**: Well-known petrochemical player Dow has been recognized globally for its stellar ESG performance and sustainability efforts. The company has been placed in various global sustainability indices such as the Dow Jones Sustainability Index (DJSI) and the FTSE4Good Index (which assess companies based on their ESG performance). Dow has also received accolades such as the US Environmental Protection Agency (EPA) *"Energy Star Partner of the Year Award"* for its energy management initiatives [32–34].
3. **Total**: Global energy and petrochemical giant Total has been aptly acknowledged for its sustainability stewardship and remarkable ESG performance. The company has received accolades such as the CDP Climate Change-A List and the RobecoSAM Sustainability Award, as a global recognition for businesses for their initiatives towards climate change mitigation, environmental advocacy, and community development initiatives [35].
4. **ExxonMobil**: Even though oil and petrochemical companies work in challenging circumstances, global energy, and petrochemical giant, ExxonMobil has been recognized for its ESG efforts. The company finds a place in sustainability indices such as the DJSI North America and the Carbon Disclosure Project (CDP) A List for climate change disclosure and sustainable performance [36].
5. **Shell**: Like its counterparts, Shell has also been rewarded for its exemplary leadership in sustainability and the ESG framework. The company has been conferred accolades such as the Carbon Trust Standard for Carbon—which recognizes organizations for their efforts to *measure, manage, and reduce carbon emissions*. Shell has also found a place in prominent global sustainability indices such as the FTSE4Good Index and the Euronext Vigeo World 120 Index [37].

The above cases exemplify the approbation the petrochemical companies receive towards their best HSE management practices and their stellar performance on the ESG compliance framework. Such awards, while motivating the petrochemical companies, also help towards a good perception of these companies.

5.13 CONCLUSION

The significance of HSE management and ESG compliance in the petrochemicals sector is acknowledged by all stakeholders of the industry. That's because any system ensuring the safety and well-being of employees and the local community is well appreciated. Also, minimization of (negative) environmental impact, mitigation of operational risks provide the petrochemical companies, a competitive edge. Similarly an effective HSE management contributes towards accident-prevention by reducing the chances of any potential liabilities for petrochemical companies. Similarly, ESG compliance enables the industry to be in tune with the requirements of modern-day stakeholders including investors, governments, shareholders, communities, and the customers. Taken together, HSE management and the following ESG norms are pivotal to the petrochemical industry's long-term business viability, positive public perception, and a competitive edge in the market in these competitive times.

NOTE

1 This discussion is from an academic perspective and in no way a judgment on the concerned company's competence to handle such incidents.

REFERENCES

1. Kok, T. L., Choong, Y. J., Looi, C. K., & Siow, J. H. (2019). Bhopal gas tragedy – the scar of process safety. *Loss Prevention Bulletin,* 269. www.icheme.org/media/12469/sieso-2019.pdf
2. Reid, M. (2020). The Piper Alpha disaster: A personal perspective with transferrable lessons on the long-term moral impact of safety failures. *ACS Chemical Health & Safety, 27*(2), 88–95. https://doi.org/10.1021/acs.chas.9b00022
3. Gurung, K., Jayadeep, L., Siwek, J., Vora, S., & Zhou, D. (2020). Texas City Refinery explosion — safety out of focus. *Loss Prevention Bulletin,* 275. Retrieved May 20, 2024 from www.icheme.org/media/14611/lpb275_pg04.pdf
4. United States Government. (2011). Report of the [Deep Water: The Gulf Oil Disaster and the Future of Offshore Drilling] National Commission on the BP Deepwater Horizon Oil Spill and Offshore Drilling (ISBN: 978-0-16-087371-3). Retrieved May 20, 2024 from www.govinfo.gov/content/pkg/GPO-OILCOMMISSION/pdf/GPO-OILCOMMISSION.pdf
5. Yu, M., Lv, Q., Ding, H., Zeng, X., Cao, J., Liu, J., Fan, H., & Hou, S. (2016). Evaluation of blast injury patients from the 2015 Tianjin explosions in China. *Burns: Journal of the International Society for Burn Injuries, 42*. doi: 10.1016/j.burns.2016.03.004.
6. HAZOP. A study for Mangalore Refinery and Petrochemicals Limited (MRPL) Aromatic Complex. (n.d.). Retrieved May 20, 2024 from https://ifluids.com/hazard-and-operability-hazopstudy-for-mangalore-refinery-and-petro-chemicals-limited-mrpl-aromatic-complex/
7. Mangalore Refinery & Petrochemicals Ltd. (n.d.). www.mrpl.co.in/sites/default/files/4517147%20MANGALORE%20REFINERY%20%26%20PETROCHEMICALS%20LTD_0.pdf
8. TotalEnergies. Certificates & documents. (n.d.). Retrieved May 20, 2024 from https://polymers.totalenergies.com/certificates-documents
9. Haldia Petrochemicals. Haldia Petrochemicals becomes India's First with BSI ISO 27001:2022 Certificate. (n.d.). TNPSCTherVupettagam. Retrieved May 20, 2024

from www.tnpscthervupettagam.com/currentaffairs-detail/bsi-iso-27001-2022-certificate#:~:text=Haldia%20Petrochemicals%20becomes%20India's%20First,program%20published%20in%20October%202022.

10. SIPCHEM. Management system certificate. Retrieved May 20, 2024 from www.sipchem.com/sites/default/files/2022-01/ISO%2027001%20-%20SIPCHEM%20-%20RC.pdf
11. OPaL India. Financial results. (n.d.). Retrieved May 20, 2024 from www.opalindia.in/financial-results.html
12. OPaL India. About Us. (n.d.). Retrieved May 20, 2024 from www.opalindia.in/aboutus.html
13. OPaL India. IMS policy. (n.d.). Retrieved May 20, 2024 from www.opalindia.in/ims-policy.html
14. OPaL India. Annual report. (n.d.). Retrieved May 20, 2024 from www.opalindia.in/annual-report.html
15. OPaL India. Linkedin post on safety management. Retrieved May 20, 2024 from www.linkedin.com/posts/opal-india_fireserviceweek-safetyfirst-opal-activity-7186687932055896064-K9gf
16. OPaL India. ICC Environment Excellence Certificate. (n.d.). Retrieved May 20, 2024 from https://opalindia.in/assets/pdf/awards-recognitions/OPaL_received_Certificate_of_Merit_from_ICC.pdf
17. Saudi Aramco. Workforce health and safety report. (n.d.). Retrieved May 20, 2024 from www.aramco.com/en/sustainability/responsible-business/workforce-health-and-safety
18. Shell. Safety approach. (n.d.) Retrieved May 20, 2024 from www.shell.com/sustainability/safety/our-approach.html
19. ExxonMobil. Sustainability report. (n.d.). Retrieved May 20, 2024 from https://corporate.exxonmobil.com/sustainability-and-reports/sustainability/caring-for-our-workforce/enhancing-process-safety
20. Dow. ESG report. (2021). Retrieved May 20, 2024 from https://corporate.dow.com/content/dam/corp/documents/about/066-00397-01-2021-esg-report.pdf
21. Dow. HSE policy. (2023). Retrieved May 20, 2024 from https://corporate.dow.com/en-us/about/legal/public-policy/environment-health-safety.html
22. TotalEnergies. Safety health environment quality charter. (2021). Retrieved May 20, 2024 from https://totalenergies.com/sites/g/files/nytnzq121/files/documents/2022-04/te_charte_hseq_en_09_21.pdf
23. SABIC. Annual report. (2022). Retrieved May 20, 2024 fromwww.sabic.com/en/reports/annual-2022/corporate/esg
24. SABIC. Climate, energy and resource efficiency. (n.d.). Retrieved May 20, 2024 from www.sabic.com/en/sustainability/climate-energy-resource-efficiency
25. SABIC. Circular economy. (n.d.). Retrieved May 20, 2024 from www.sabic.com/en/reports/sustainability-2021/innovation/circular-economy
26. SABIC. Safety and health. (n.d.). Retrieved May 20, 2024 from www.sabic.com/en/about/ehss/safety-and-health#:~:text=SABIC%20and%20its%20affiliates%20are,and%20contractor%20injuries%20and%20illnesses
27. SABIC. Social impacts and community relationships. (n.d.). Retrieved May 20, 2024 from www.sabic.com/en/reports/sustainability-2020/collaboration/social-impacts-and-community-relationships

28. SABIC. Corporate governance. (n.d.). Retrieved May 20, 2024 from www.sabic.com/en/investors/corporate-governance
29. SABIC. ESG. (n.d.) Retrieved May 20, 2024 from www.sabic.com/en/reports/annual-2022/corporate/esg
30. SABIC. Innovation and collaboration. (n.d.). Retrieved May 20, 2024 from www.sabic.com/en/news/43030-sabic-supports-industry-transformation-through-innovation-and-collaboration
31. Saudi Aramco. Green World Awards. (n.d.). Retrieved May 20, 2024 from www.aramcolife.com/en/publications/the-arabian-sun/articles/2023/week-23-articles/aramco-earns-11-green-world-awards
32. Dow. Sustainability Awards. (n.d.). Retrieved May 20, 2024 from https://corporate.dow.com/en-us/news/press-releases/dow-recognized-with-four-sustainability-awards.html
33. Dow. Chemical Week Sustainability Awards. (n.d.). Retrieved May 20, 2024 from www.3blmedia.com/news/dow-named-finalist-chemical-week-inaugural-sustainability-awards
34. Dow. Energy and waste management award. (n.d.). Retrieved May 20, 2024 from https://corporate.dow.com/en-us/seek-together/dow-recognized-for-sustainability-efforts.html
35. TotalEnergies. Main ESG ratings. (n.d.). Retrieved May 20, 2024 from https://totalenergies.com/sustainability/reports-and-indicators/main-esg-ratings
36. ExxonMobil. Responsible Care Company of the year Award. Retrieved May 20, 2024 from www.prnewswire.com/news-releases/exxonmobil-receives-responsible-care-company-of-the-year-award-for-second-consecutive-year-301844268.html
37. Shell. Sustainability report 2022. (2023). Retrieved May 20, 2024 from www.knowesg.com/investors/shell-releases-2022-sustainability-report-29032023

Disclaimer: All brand logos, trademarks, company names, company photographs, and company data mentioned in this chapter are the intellectual property of their respective owners. These are used for informational and educational purposes only and are not intended to imply any endorsement or affiliation with the brands/companies. The inclusion of this information is for illustrative purposes to support the analysis and discussion in this academic work. No commercial use or exploitation of the logos, trademarks, company names, company photographs, and company data is intended or should be inferred.

Index

D

Q

R

T

U

V

W

X

Y

Z

Made in the USA
Las Vegas, NV
19 September 2026

d5177c6c-33a8-441c-a9c5-c161a7993aa9R01